Europe from Crust to Core

Europe
from
Crust to Core

Europe

from

Crust to Core

Keynote Addresses from the First Meeting of European Geological Societies,
Reading (UK), September 1975

Edited by

D. V. Ager

and

M. Brooks

Department of Geology
University College of Swansea

A Wiley–Interscience Publication

JOHN WILEY & SONS
London · New York · Sydney · Toronto

Library of Congress Cataloging in Publication Data:

Meeting of European Geological Societies, 1st, Reading, Eng., 1975.
　Europe from crust to core.

　'A Wiley–Interscience publication.'
　Includes index.
　1. Geology—Europe—Congresses. I. Ager, Derek Victor. II. Brooks, M. III. Title.

QE260.M4 1975　　554　　76-40096
ISBN 0 471 99420 0

Printed in Great Britain at The Spottiswoode Ballantyne Press,
by William Clowes and Sons Ltd., London, Colchester and Beccles

PREFACE

Though one of the smallest continents, Europe is arguably the most complex politically, linguistically and culturally. It is therefore all the more welcome that there are increasing signs of significant and lasting contacts between its diverse peoples, not this time through war and military conquest but through trading agreements, cultural exchanges, popular travel, the influence of the mass media, and scientific and technical co-operation. European earth scientists have been at the forefront in the field of international scientific endeavour.

The early growth of geology was much influenced, of course, by international communication, consort and travel. One thinks of the long and acrimonious debate between Neptunists and Plutonists, taking place across frontiers in an eighteenth-century Europe torn by war and revolution. Lyell's *Principles of Geology* is illuminated by his wide experience based on travels in many European countries. International co-operation in the earth sciences, however, was perhaps born in Europe in the late nineteenth century and contributed to the monumental achievements of generations of Alpine geologists. Notwithstanding the early importance of national schools, Alpine geology has always been characterized by international co-operation and one may point to the international explosion seismological experiment of 1975 (ALP 75) as the latest example of this long tradition.

Nonetheless, until quite recently geology in most European countries, and perhaps particularly in Britain, was very much a domestic pursuit based on text books and field work largely unconcerned with geology beyond national frontiers. Even where geologists have worked in foreign countries they have sometimes retained a parochial outlook so that, for example, the tendency for geological boundaries to coincide with national boundaries has not been successfully overcome.

The recent revolution within the earth sciences has done much to broaden outlooks and encourage international co-operation. Plate tectonic theory highlights the large scale, as well as the multi-disciplinary nature, of geological problems. Consequently we can no longer delude ourselves that pottering around in our backyards (any more than considering a narrow specialist field in isolation) will lead to an adequate understanding of major earth processes and planetary evolution.

The British have tended to think of themselves as rather separate from the rest of the continent, by reason of their geography and history (both of course very much controlled by the geology), so perhaps the realization of the unity of Europe came later to them than to most. However, it is well known that none are stronger in their beliefs than the converted and early in the 1970s a few British geologists got together to plan the first Meeting of European Geological Societies to be held in Reading in September 1975. It was held under the auspices of the Geological Society of London whose motto is very appropriate in this context: *Quicquid sub terra est.* The intention was to consider the geology of Europe in its true and fullest sense. Nearly 400 people attended that meeting and represented 35 countries; only three major European nations were unrepresented. It was decided not to follow the muddy path so frequently trodden by scientific meetings. Papers were not invited. Professor Percy Allen proposed the main theme of 'Europe from Crust to Core' and this was divided into a series of lesser themes for each of which a 'Keynote' speaker was selected. These were invited to give stimulating and provocative addresses which would lead to lively informal discussions. This volume contains most of those keynote addresses. The rest of the conference was taken up with short communications related to the main themes, and these were distributed in abstract form, but could not be included in the present volume. It is also very much regretted that, for various reasons, the keynote addresses on applied geology in Europe, which led to some of the liveliest discussions, and the proposed sessions on European palaeontology are not included here. It was a particular sorrow to us that, at the last moment, Professor Rudolph Trumpy of Switzerland was prevented by illness from coming to give his address on the western part of the Alpine system.

The main divisions of the meeting and of this book are those great divisions of Europe recognized by Stille 50 years ago and still applicable in this age of plate tectonics.

The original intention of the conference – as enunciated by Professor Allen – was to consider 'Europe from Crust to Core', and that is also the title of this book, but the conference was not wholly concerned with European matters. As Professor Jacobs' chapter makes clear, the European part of the core cannot be considered separately from that of the rest of the world. Similarly there are obviously aspects of geological investigation which, though largely carried out by European geologists and based on European examples, are world-wide in application. Professor Ramberg's contribution is very much in this category.

It is always difficult to transform the spoken into the written word and we thank all the contributors for their tolerance of the way we have treated their texts. We also apologize to all our continental colleagues for our irritating insularity in presenting this book entirely in the English language.

Special thanks are due to Mrs. Jose Nuttall, Mrs. Jan Greengo and Miss Avril Davies for their help in preparing the final text of this volume and to Mr. John Edwards for his work on several of the illustrations.

Swansea
January, 1976

D. V. AGER
M. BROOKS

CONTRIBUTING AUTHORS

PROFESSOR P. ALLEN, Department of Geology, University of Reading, Whiteknights, Reading, RG6 2AB, England

PROFESSOR J. A. AUBOUIN, Département de Géologie Structurale, Université Pierre et Marie Curie, 4 Place Jussieu, 75230 Paris, France

SIR KINGSLEY DUNHAM, F.R.S., Institute of Geological Sciences, Exhibition Road, South Kensington, London SW7 2DE, England

DR J. M. HARRISON, Assistant Director-General for Science, UNESCO, 7 Place de Fontenoy, 75700 Paris, France

PROFESSOR J. A. JACOBS, Department of Geodesy & Geophysics, University of Cambridge, Madingley Rise, Madingley Road, Cambridge, CB3 DEZ, England

PROFESSOR V. E. KHAIN, Sub-Commission of the Tectonic Map of the World, 109017, Moscow 17, Pyjevski per. 7, USSR

PROFESSOR W. KREBS, Institut für Geologie und Paläontologie der Technischen Universität, D-33 Braunschweig, Pockelstrasse 4, Federal Republic of Germany

PROFESSOR A. KVALE, Department of Geology, Universitat i Bergen, Bergen, Norway

DR G. LÜTTIG, Bundesanstalt für Geowissenschaften und Rohstoffe und des Niedersachsischen Landesamtes für Bodenforschung, 3 Hannover-Buchholz, Stilleweg 2, Federal Republic of Germany

PROFESSOR G. F. MITCHELL, F.R.S., Department of Geology, Trinity College, Dublin 2, Eire

PROFESSOR H. RAMBERG, Institute of Geology, University of Uppsala, Uppsala 1, Sweden

PROFESSOR J. WATSON, Department of Geology, Imperial College, London S.W.7, England

CONTENTS

Introduction
Earth Science in Europe
  Sir Kingsley Dunham 3
European Geology—the Future
  P. Allen 12

The General Framework
The New International Tectonic Map of Europe
  V. E. Khain 19
The Earth's Deep Mantle and Core
  J. A. Jacobs 41

Eo-Europa
The Evolution of a Craton
  J. Watson 59

Palaeo-Europa
Major Features of the European Caledonides and their Development
  A. Kvale 81

Meso-Europa
The Tectonic Evolution of Variscan *Meso-Europa*
  W. Krebs 119

Neo-Europa
Alpine Tectonics and Plate Tectonics: Thoughts about the Mediterranean
  J. Aubouin 143
Some Models Illustrating Tectonic and other Processes in the Lithosphere
and Upper Mantle
  H. Ramberg 159

Quaternary Europa
Problems and Aims in Quaternary Europe
  G. F. Mitchell 169

The Future of European Geology
Organization of European Geology: Present and Future
  J. M. Harrison 183
The Role of Geosciences in Modern Society
  G. Lüttig 192

Index 199

Introduction

EARTH SCIENCE IN EUROPE

SIR KINGSLEY DUNHAM

Institute of Geological Sciences, London, UK

Abstract

Dunham, K. (1976). 'Earth Science in Europe, in Ager, D. V. and Brooks, M. (eds.), *Europe from Crust to Core*, Wiley, London.
The need for a greater degree of collaboration in the earth sciences in Europe is discussed. Close relationships between individuals have existed for two centuries, but European organizations to promote collaboration are in their infancy. The international unions (IUGG and IUGS) do something towards this objective. A European Geophysical Society has been founded. The new European Science Foundation based at Strasbourg will eventually become influential in Western Europe. A network of bilateral exchange agreements of which the Royal Society European Programme is an example facilitate the logistics of collaboration. Some aspects of the earth sciences may not yet be adequately provided for.

Historical Perspective

Without wishing to be accused of chauvinism on a continental scale, I must begin by claiming that Europe was for many years the cradle of geology. Classical Greece gave us Xenophanes of Collophon, Aristotle, Theophrastus, Anaximander, Empedocles of Agrigentum, Strabo and two score other inquirers into the story of the earth, not all of whose works have survived. From ancient Rome we remember not only Pliny the Elder, whose enthusiasm for observing volcanism at close range cost him his life at Vesuvius, but also Lucretius' *De Rereum Natura* and the physician Galen. The contributions from this period may have been, as Adams (1938) says, 'scanty and of little worth', but they include the first writings that indicate the application of the human mind to the problems of the European crust and seas. The word *geologia* was first coined by a fifteenth-century bibliophile Bishop of Durham (Bury, 1473), but in its modern sense it appeared two centuries later.

In later mediaeval time, the growth of a metal-mining industry, especially in central Europe, stimulated knowledge of mineralogy and the practical arts, splendidly summarized by the sixteenth-century writer Agricola (1556) while the Renaissance in Italy brought the universal mind of Leonardo (Anon, 1964) to bear on the problem of fossils; for some of us, he is the real founder of our subject. The discoveries made during the mining of the stratiform copper deposit of the Kupferschiefer around Mansfeld near the Harz Mountains were fundamental: by the time of Lehmann and von Justi no less than 49 different layers (i.e. formations) had been recognized in the district. It was indeed the contrast between

3

the contorted slates and crystalline rocks of the Harz and the layered 'secondary' rocks of the plain to the south that formed the basis for Werner's brilliant teaching at Freiberg, and the rise of the Neptunists. Meanwhile in St. Petersburg, the father of Russian geology, Lomosonov was active. The disciples of Werner spread his doctrine far and wide in Europe, provoking in due course a brisk reaction to their over-insistence on the role of oceanic sedimentation from James Hutton of Edinburgh. He became the leader of the Plutonists who believed in a deep-seated, molten origin for igneous rocks and could offer experimental evidence in support of their views. It is no part of mine to attempt to trace the great and bitter controversy between the Neptunists and the Plutonists except to express the belief that the theses advocated by both sides were a vital ingredient in the progress of our subject, which thrives as I see it on controversy. Nor need I trace the rise of modern earth science through Cuvier, William Smith, Charles Lyell, Darwin, Kelvin, Rosenbusch, Lacroix, Bertrand, Suess and the rest to modern tectonics, geophysics, geochemistry, palaeoecology; it is enough to say that we in Europe share in a rich heritage of history. It would be pleasant to pursue this theme and to develop a Valhalla of geological heroes to which every European country could contribute several, but our business here in Reading is with the future rather than the past. We may nevertheless notice that a great deal of international travel and collaboration went on from the late eighteenth century in pursuit of the earth sciences. To take only one example, Charles Lyell (the centenary of whose death has just been celebrated in London and at Kinnordy by the International Committee on the History of Geology), travelled extensively in Europe, visiting and becoming well-known to the leading geological savants in France, Italy, Germany, Switzerland, Sweden and Greece. The subdivision of the Tertiary, which we continue to use, was as much or more influenced by Italian material as by collecting in this country.

Geological Surveys

The work of the European pioneers led, during the nineteenth century, to a widespread recognition by governments that it would be valuable to have the outcropping rocks and superficial deposits recorded on maps. Much of the continent and adjacent islands has been under official geological survey for more than a hundred years. Many of the organizations concerned have celebrated their centenaries in my lifetime. The Geological Survey of Great Britain (now part of the Institute of Geological Sciences) was first in the field in 1835, though there had been earlier mineral surveys in France and soil surveys in England. The Service de la Carte Géologique de France was begun in 1869. It now forms part of the Bureau de Recherche Géologique et Miniéré. The Österreichische Geologishe Bundesanstalt is 125 years old this week, and the Bayerische Geologische Landesamt is 125 years old on 15 September. The centenary of the Hungarian Geological Institute was celebrated in 1969, and that of the Koenigliche Prussiche Geologische Landesamt (now the Bundesanstalt für Geowissenschaften und Rohstoffe (BGR)) in 1971. These are only a few examples among many. I have

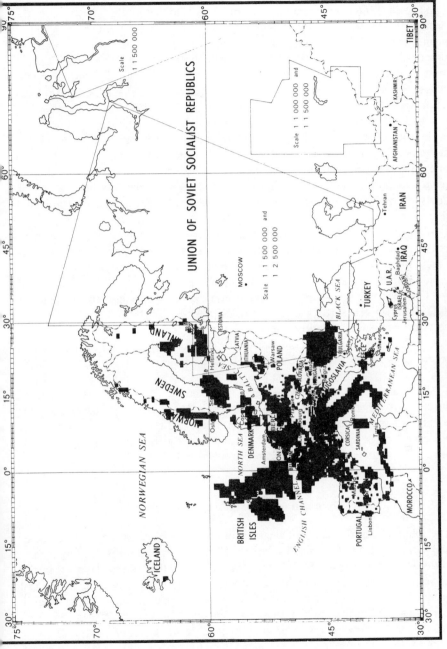

Mapping at a scale of 1:100 000 and larger (except U.S.S.R. mapped at scales shown above)

Figure 1

claimed (Dunham, 1967) that Europe is the most intensively mapped part of the globe: Fig. 1 illustrates this contention though based on information available in London in 1966, and admittedly incomplete. Thanks to the activities of the Secretary-General of the International Union of Geological Sciences, most European Geological Surveys have provided up-to-date statements of their mapping progress for publication in the *Geological Newsletter* and these show that the coverage of geological maps is still more extensive in Europe than my diagrams suggest. (The full list of summaries of published geological maps for 27 European countries is given in the *Geological Newsletter*, 1975, **2**, 198.) In all countries, the key geological areas have been revised several times over, in the light of new subsurface information and new geological techniques. Nevertheless, I am certain, as Director of a geological survey, that information now pouring in as a result of the intensified search for fuel, water, high-cost minerals and bulk minerals will continue to make further revisions imperative, and probably the development of new kinds of geological cartography, for example geotechnical and environmental maps.

With so many field data available, at least two alternative points of view may be adopted. One is to say that earth science in Europe is within sight of being finished. This is the view one may expect to hear from people not familiar with the rapid technical progress in the subject. I recall that at the time of the centenary of the GSGB, forty years ago, a London evening newspaper carried the headline 'GOVERNMENT SURVEY 100 YEARS OLD, STILL NOT FINISHED'. But very few geologists will subscribe to the view that earth science in this continent has no future. On the contrary, the wealth of accurately located data should be regarded as one of our greatest assets, both as a basis for theoretical investigations and as essential background for all kinds of practical activities.

The question to be asked—and I shall leave it to Professor Allen to develop—is whether we are making sufficient use of the opportunities for collaborative examination of this great data-base in Europe. Before he does that, however, I propose to review briefly the nature of the international organizations already promoting geological collaboration in and beyond Europe.

International Congresses and Unions

The International Geological Congress, one of the first scientific conferences of its kind, met for the first time in Paris in 1879, having been organized by a committee under the presidency of James Hall (USA), with T. Sterry Hunt (Canada) as secretary, and dominated by representatives of the New World. It has met generally at four-year intervals, and of its 24 meetings so far, 15 have been in Europe, six in North America, two in Africa, one in Asia. The 25th session next year will be the first in Australia. In each case, the whole weight of organizing the congress has fallen upon the host country, and the amount of collaborative effort, though undoubtedly stimulated by the meetings, has fallen off to a marked extent between sessions.

The movement which began after World War I to create scientific unions, linked with the International Council of Scientific Unions, included as one of the first four the International Union of Geodesy and Geophysics in 1919. This links six major international associations (Geodesy, Seismology and Physics of the Earth's Interior, Volcanology and Chemistry of the Earth's Interior, Geomagnetism and Aeronomy, Meteorology and Atmospheric Physics, Hydrological Sciences, Physical Sciences of the Ocean). Through the years this organization has played a most significant part in bringing together physical and earth scientists. The membership includes 19 European countries out of a total of 71 nations, but these figures do not accurately reflect the part played by Europeans in the Union's work.

For those parts of geology not directly linked with physics, the International Congress was held to be a sufficient forum up to the time of the Copenhagen meetings of the Congress in 1960. It was then felt that, to provide continuity between congresses and to achieve a linkage with ICSU and UNESCO, the formation of an International Union of Geological Sciences was desirable. This was launched in Paris in 1961, its Council first met in Rome in 1963 and by the time of the Delhi International Congress in 1964 it was fully operational. The commissions, including the Commission on Stratigraphy, were transferred from the Congress to the Union and there are now numerous affiliated international associations. A total of 76 countries have joined the union, of which 23 are in Europe, and this continent has so far provided two out of four presidents.

Perhaps the greatest achievement of the IUGS so far has been the organization of the first joint project between a scientific union and UNESCO. Known as the International Geological Correlation Programme, the concept, which first emerged in 1964, was elaborated at Prague in 1967 and 1968, discussed by a representative gathering in Budapest in 1969, further developed in Paris in 1971, and its constitution was adopted in Toronto in 1972. It is fair to say that it could be far more effective had it ample funds with which to pay for research on correlation across national boundaries; but it is already acting as a useful catalyst in providing the means to hold organizing meetings for its adopted projects. National participation, however, mainly depends on national funding. Priority is deliberately being given to the two ends of the geological time scale, that is to the Precambrian and the Quaternary, and special emphasis is being laid on accuracy in time determination, Phanerozoic events being by no means excluded. As the programme stands at present, Europe provides leadership for 16 out of the 21 adopted key projects within the programme.

It should not be supposed from what has been said that IUGG and IUGS pursue wholly separate courses. Apart from attendance by physically and non-physically inclined geologists at meetings of both, a major area of mutual interest has been developed in the Inter-union Commission on Geodynamics (Sutton *et al.*, 1975), which succeeded the very successful Upper Mantle Project. This divides into nine subject areas, and though it is due to terminate in 1978, the research impetus it has created in global tectonics and many other fields will certainly continue.

The unions and inter-disciplinary projects each have a national committee structure behind them. In some countries the national academy of sciences (in Britain, the Royal Society) provides this structure; in others it is linked to the geological surveys.

I have felt it right to include this brief summary of the international organizations since at the Reading meeting we must clearly consider whether or not a continuing pan-European organization needs to be set up, or whether the existing international machine is sufficient to promote a healthy science of the earth in Europe.

European Science Foundation

It is next necessary to refer to purely European organizations. The newly formed Science Foundation calls for some comment first.

In 1972 Commissioner Astiero Spinelli (1972) published proposals for a common scientific, technological and industrial policy for the European Economic Community. These proposals appeared to be liable to lead to the setting up of a bureaucracy not only to foster the politically sensitive development aspect of R & D but to attempt to run fundamental scientific research. They were considered by a meeting of representatives of the learned academies of Europe at the Royal Society on 1 December of the same year from which it emerged (i) that any such foundation should separate fundamental from applied research (ii) that it would best be organized from below, by scientists themselves and (iii) that the membership ought not to be limited to the nine members of the EEC. It was also noted that an association of European research councils, the so-called Aarhus Group, had begun to meet. At Munich in April 1973, Dr. R. Dahrendorf, now Commissioner responsible for European science, met the academicians and gave considerable support to their ideas for a European Fundamental Science Foundation, which he further amplified at the next meeting, held in Gif-sur-Yvette in September. Here the decision in principle to set up the Foundation was taken; at Saltsjöbaden, near Stockholm, in May 1974 a decision was taken to base the Foundation at Strasbourg in premises provided by the city and with the assistance of the Centre National de Recherche Scientifique of France. The founding meeting, at which the constitution was ratified, took place at Strasbourg on 18 November 1974, less than two years after the initial discussion in London. Welcome and encouragement from the European Commission was expressed by Commissioner Brunner who by now had succeeded Dr. Dahrendorf. Sir Brian Flowers (UK) was elected president; Professor O. Reverdin (Switzerland) and Dr. P. Riis (Denmark) vice-presidents; Dr. F. Schneider (West Germany) Secretary-General. The Council consists of one member from each member-country, the list including Austria, Belgium, Denmark, France, W. Germany, Greece, Italy, Ireland, Netherlands, Norway, Portugal, Spain, Sweden, Switzerland, United Kingdom, Yugoslavia. The present representative of Switzerland, Dr. P. Fricker, is a geologist and as far as possible a balance of disciplines, which cover not only the physical and biological sciences, but also medicine and the humanities, has been attempted. The science and medical research councils of Europe, and those academies that finance research, will be fully represented at Assembly meetings.

One further point is of some importance. The Foundation, though initially West European, is not in the long run intended to be an exclusively western association and many members hope that in due course it will become pan-European in scope.

Its object, besides providing a forum, is to encourage projects in basic science and learning which cannot be carried out by any country individually from its own resources. There are many such projects in our field, and it is very desirable that earth sciences should play a full part in this interesting manifestation of Europeanism.

European Societies

In the practical field, there are two societies which deserve mention. The European Society of Exploration Geophysicists, formed after World War II, meets annually in a European capital, and attracts industrial, government and consulting geophysicists and makers of equipment. The Society for Geology Applied to Mineral Deposits (SGA) came into existence in 1965, to meet a need for publication in European languages other than English (though it also publishes in English). The Society of Economic Geologists, based in the USA, did not feel able to undertake this, and the success of the SGA and its journal *Mineralium Deposita* has shown that the need for a European forum certainly existed. The society is, however, by no means confined to Europe in its interest and may resent my including it here; but it had a truly European origin in which G. C. Amstutz, G. L. Kroll, A. Bernard, A. Maucher and particularly P. Ramdohr played leading parts. The SGA is linked with SEG and the International Association on the Genesis of Ore Deposits in the International Federation of Societies of Economic Geology, and so with IUGS.

More theoretical geophysics is the field of a newer society, the European Geophysical Society, proposals for which originated at the joint meeting of the International Associations of Geomagnetism and of Seismology in Madrid (Anon., 1973) in 1969. The foundation was furthered at a meeting in Reading in 1971 where Professors Runcorn and Allen were among the sponsors, and Mr. C. R. Argent of the Royal Society became secretary. The first full meeting, at Zürich in September 1973, attracted 500 participants.

It is probably fair to say that all three societies have to some extent been influenced by pan-American models, for example the American Geophysical Union, an organization so successful that a decade ago its secretary-general was claiming that it had 'taken over geology'. The claim was not one that appeared to many non-physics-based geologists, but the idea behind it that some attempt should be made to see earth science as an integrated whole is praiseworthy enough.

Of the national geological, palaeontological, mineralogical, petrological, pedological, sedimentological, geophysical, geochemical and related societies, I need only say that they are, and may be expected to remain, the main organs for discussion and publication and, hence, powerful instruments for encouraging research. Few of us, I imagine, believe that we have yet reached a stage where the identities of these societies should be merged in pan-European associations; probably we are as far from that as from the formation of one, or perhaps two, United States of Europe. Nevertheless, there is a possible intermediate position, as represented by the truly European societies. Ought collaboration to be strengthened by promoting another organization, one with still wider terms of reference?

Exchange of Scientists

Perhaps the most effective way of ensuring understanding between scientists and collaboration across national frontiers is to make it financially possible to exchange scientific workers; indeed one might go further and say that unless this is done, real collaboration will be difficult to achieve.

The past decade has witnessed a remarkable growth in bilateral agreements between academies and research councils in Europe with this objective. The Royal Society's European Programme, launched in 1964, is based upon balancing financial arrangements whereby both parties to a bilateral agreement make available equivalent funds. With the conclusion, expected shortly, of such an agreement with the Yugoslavian Federation of Academies, the Royal Society will have agreements with academies or research councils in all European countries save two. The scheme makes possible in both directions study visits of a few weeks, or fellowships of from six months to two years duration. Geologists have benefited from the scheme, along with scientists of all other disciplines. Comparable exchange arrangements are operated by a number of other organizations; for example, the Centre National de Recherche Scientifique, the Deutsches Forschungsgemeinschaft, the Alexander Humbolt Stiftung, the Polish Academy of Sciences, to mention only a few. Of more narrow scope is the fellowship arrangement associated with the North Atlantic Treaty Organization.

In the end, however, the financing of scientific collaboration in research comes back, in most cases, to government money, as indeed does the support for a high proportion of all research. We may therefore usefully ask ourselves, will the kinds of research exposed at the present meeting convince governments that they ought to be supported and encouraged on a European scale?

The Tasks Ahead

I would not presume to attempt to answer this question, but it may be worthwhile to mention a few topics for collaborative geological research in Europe which are, in my judgment, of both practical and high theoretical interest:

(1) The proper understanding of the superficial unconsolidated deposits. The fundamental importance of these in geotechnics is by now well appreciated; but

by no means all the problems are solved by measuring physical properties. In particular, much more attention to the (admittedly difficult) stratigraphical problems involved is clearly needed.

(2) The relationship of active tectonics, global and local to these deposits, and to areas of recent and active seismicity and volcanism.

(3) The geothermal gradient in the top 4–5 km of the European crust, its relationship to stratigraphy, igneous rocks, metamorphic rocks, thermal waters, mineralization.

(4) Phanerozoic stratigraphy, palaeotectonics and palaeogeography of Europe. A great deal is known after 150 years of close investigation, yet many problems of correlation across national boundaries remain. The relationship of these to fossil fuel (petroleum, coal, lignite) and mineral fuel (uranium ore).

(5) The nature of the lower crust beneath Europe and the question as to whether the exposed Precambrian of Scandinavia/Scotland, the Ukraine and elsewhere is representative of this part of the crust where a thick sedimentary cover exists, or under the great orogenic belts.

(6) The geology, in its widest sense, of the Mediterranean, the European continential shelves, and the Atlantic continental margin.

References

Adams, F. D. (1938). *The Birth and Development of the Geological Sciences*, Williams & Williams, Baltimore

Agricola, G. (G. Bauer) (1556). *De Re Metallica*, Basel

Anon. (Inst. Geog. Agostini) (1964). *Leonardo da Vinci*, Novara

Anon. (1973). 'European science foundation formed', *Nature*, **245**, 228

Bury, R. de (1473). *Philobiblon*, Koln

Dunham, K. C. (1967). 'Practical geology and the natural environment of Man—I. Continents and islands', *Quart. J. geol. Soc., Lond.*, **123**, 1

International Geological Correlation Programme (1975). Third session of the (I.G.C.P.) Board, *Geol. Correlation*, **3**, (*in the press*)

Spinelli, A. (1972). *Objectives and instruments of a common policy for scientific research and technological development*, European Economic Commission, Bruxelles

Sutton, J. *et al.* (1975). *Geodynamics Today. A Review of the Earth's Dynamic Processes*, Royal Soc., London

EUROPEAN GEOLOGY—THE FUTURE

P. ALLEN

Department of Geology, University, Whiteknights Park, Reading, UK

MEGS Objectives

In providing a forum for discussing the *integrated geology of Europe as a whole*, MEGS also prompted the questions: (i) What are Europe's geological and sociogeological tasks and responsibilities and (ii) How efficient are Europe's geological organization and facilities for carrying these out?

'Geology' is of course the science of the earth beneath our feet. MEGS used the word in this sense, though some might have preferred 'geoscience' (which is plainer) or 'geological sciences' (which is repetitive and longer). Let us be clear on one issue, however: our subject comprises a fabric of geobiology, geochemistry and geophysics, pervaded where appropriate by mathematics. (There is nothing else but the human spirit.)

European Tasks

What are our continental responsibilities? Europe lies on a gigantic rock, gas and liquid cone, with its apex at the Earth's centre. Sir Kingsley Dunham has reminded us that it is the most intensively studied segment of our planet.

Undoubtedly our basic task is to determine the structure and processes of this 'Eurocone', now and over the past four and a half billion years, in order to

(1) use it as an 'experimental animal' for discovering principles applicable to the world generally;

(2) exploit its resources for human good;

(3) manage this exploitation sensibly (geology being basic to man's economy and environment);

(4) employ our resulting predictive powers for preserving Europe's environment and quality of life;

(5) use the knowledge obtained for assisting the 'Third World'.

You will note that I have said nothing about the *control* of geological processes. This requires a measure of pan-European wisdom, a subject beyond the scope of MEGS.

12

European Tools

Are we facing up to our responsibilities? Have we the tools, including the organization, for our tasks? Are eurogeologists thinking *collaboratively* about continental matters? I believe the answer is 'no' on most counts, and I propose to suggest some of the things that are wrong and what is needed to put them right.

In the first place eurogeologists are over-concerned with specialist and local trivia. We all know, and pay lip service to, the dangers of this. No mantle-geophysicist can arrive at geological reality without the petrologist. Nor—dare I say it?—*vice versa*. No Quaternary chronologist ignores the core-geophysicist except at his peril. Nor the latter the former. No stratigrapher can safely correlate solely within his own country. Nor, indeed, dare his country's economists safely ignore him.

How are such barriers to be broken? Sir Kingsley pointed out that more movement of productive workers, especially younger geologists, is needed. Personal contacts and the easy forging of scientific friendships are vital ingredients of tomorrow's eurogeology. I add universal bilingualism. We need, on the continental scale, more collaborative thinking between both individuals and institutions. We need more collaborative projects, and could well take a leaf out of the oceanographers' book. Europe's travel network is dense and the distances small. Any eurogeologist or institute can be in day-to-day working contact with almost any other. More movement requires money, perhaps even MEGS to encourage it, and I shall return to these topics later. Beforehand, however, we need fundamental changes of attitude, for, as the President has reminded us, even the available money is usually not taken up. How eurogeologists are to be persuaded or enabled is less clear.

Secondly, our scientific mix is wrong. Eurogeology contains too little chemistry, too little experimental work, too little theoretical activity. Causes must be partly rooted in our educational systems, partly in job situations, partly in intellectual inertia. All three factors are interactive. With Europe's variety of peoples, educational systems and economies we may already have the means of salvation. Unleash their bonds and the resulting interplay of ideas might work wonders. But how many European universities or institutions have reciprocal staff, student or course arrangements? How many exchange staff with industry; with government scientists? Perhaps MEGS has a midwife's role here?

Thirdly, our intellectual climate is not improving fast enough. Lordgod-professors are still with us, though less chronically ill than when I was a young man. A scientist is as good as his last paper. If an eminent geologist utters claptrap then his most junior student should be able to explain why publicly, without risk to exams, career or person. This is a basic scientific freedom. Eurogeologists should be able to contact their colleagues without 'going through' administrative seniors (not necessarily betters). Personal access is another freedom essential to the efficiency and well-being of science.

Fourthly, our tools are inadequate. Geologists widely admit that Europe's vast geological data-bank is inefficiently used. Data are coming in faster than they can be processed and some institutions have already given up hope of tackling their backlogs. On occasions it is easier to repeat the original work than to dig it out of the records! Such situations are serious when the client is industrial, governmental or people in peril. Solutions lie in changing entrenched attitudes to computing and providing better facilities. But has anyone begun to think about our disgraceful computer facilities on a continental scale? Research, exploration and exploitation of Europe's natural resources are also held up on other fronts by lack of expensive modern equipment. When are we going to examine the case for setting up European centres on the physicists' models in Geneva, Strasbourg, etc? Many eurogeologists are cut off from the necessary facilities for high-temperature–pressure experimental petrology, large-scale modelling, isotopic work, sophisticated analysis and so on. Forward thinking is not being done and expertise in political pressurizing is not being acquired. Has MEGS a role to play here?

Fifthly, our organizatidnal structure is deficient. Basically good, it comprises independent local, regional, national, specialist and general societies underpinning national academies, and public and private institutes supporting governments and industries. This gives the inbuilt diversity necessary for survival. But there are gaps at the continental level, even in the specialisms. Several subdisciplines still need forums based on the European Geophysical Society model (born in the same room at Reading as MEGS) and geographical regions need forums on the Nordic model. There is also an urgent need for action at the highest level, transcending specialisms' and nations' frontiers. We need, in short, a platform for *integrated European geology.*

How this would relate to IUGS, IUGG, UNESCO, etc. I do not know. But such a forum obviously is necessary for pressurizing governments, promoting movement and collaboration and regaining our place in the forefront of science. It follows that we need more political and geopolitical skill to convince governments, funding agencies and other scientists. Geology is undoubtedly fundamental to any nation's economy and the management of its environment. But how good are we at getting the message over? How successful are we at convincing our political masters that geology is 'big science' and to be funded on that basis on a European scale? How otherwise can Europe hope to prosper and assist the developing nations? MEGS could be a 'think tank' for these problems.

Towards Eurogeoutopia

While it will be long before the preceding deficiences are successfully tackled, there is every reason for beginning to think beyond them. Whatever agencies provide funds and other support, there will always be those who make science policy and the bureaucrats who implement it. At an early date euroscientists should pool their national experiences of science funding and administration and argue out what is really required.

Our President exposed the main issue when he said, concerning the development of a European agency, that 'it would best be organized from below, by the scientists themselves'. In geology I freely translate Sir Kingsley's 'scientists' as 'active geologists, not professional geopoliticians'. The importance of this is perfectly illustrated by his history of the emerging European Science Foundation. The ESF would have been a dangerous scientific (and political) absurdity if the European scientists themselves had not forcibly intervened. In science, as in other walks of civilized life, the laboratory and field workers—men and women who actually produce the goods—are too often regarded as the 'bottom' of the system. The active European geologist should be accorded more esteem and influence than he customarily receives. Dr. P. Fricker is an important symbol as well as an important man.

The same considerations concern the 'geo' part of any European science bureaucracy and raise the old dilemma. Should such administrators be inactive or superannuated geologists exposed to the power-lusty temptation of intervening in policy matters? Or should they be safely free of original expertise in our subject?

Whatever the future of Europe, most funds will originate from governments. Two principles should be conserved in the government $\rightleftharpoons$ geology interaction. On our side we must strive to retain a strong measure of *organizational diversity* in order to mount multiple pressure groups. Proposed amalgamations of societies, institutes, etc. should be scrutinized closely in the pan-European context. Perhaps MEGS could be useful here? On the opposite side of the interaction it is desirable that the principle of *multiple funding sources* is also conserved. We all remember important advances in research, exploration and exploitation which only came to fruition because the bright young geologist by-passed the establishment or his boss. The same principle applies at 'higher' levels. But how is it to be safeguarded? Or newly established? I only know that it is a different problem for the citizens of different countries.

MEGS Programme

Pressure for some of these needs is evident in the MEGS programme: communications on a wide range of depths in the Eurocone; multi-national authorship reflecting collaborative research transcending national frontiers; a short contribution with 'European obligations' in its title; others extending the limits of Europe for geological, rather than political, reasons.

In participating in the meetings concerned with others' specialisms, in listening to tongues other than their own and in reflecting on Europe's tasks and needs, members should be becoming so integrated and Europeanized as to forgive the organizers for getting nationalities wrong—after all, they don't really matter, do they?

The General Framework

THE NEW INTERNATIONAL TECTONIC MAP OF EUROPE AND SOME PROBLEMS OF STRUCTURE AND TECTONIC HISTORY OF THE CONTINENT

V. E. KHAIN

IGC Sub-commission of the Tectonic Map of the World, 109017, Moscow 17, Pyjevski per. 7, USSR

Abstract

Khain, V. E. (1976). 'The International Tectonic Map of Europe and some Problems of Structure and Tectonic History of the Continent', in Ager, D. V. and Brooks, M. (eds.), *Europe from Crust to Core*, Wiley, London.

The second edition of the tectonic map of Europe differs from the first edition in covering not only the continent but also peripheral areas, including the surrounding sea and ocean floor. The basic subdivision, as before, is into regions defined by the age of their main deformation. Internal structures in the craton are recognized and massifs within the later fold belts are subdivided according to the age of their main deformation and degree of reworking. The main fold belts are defined in terms of their main phases of folding rather than their structural stages.

Introduction

The beginning of the international work on the Tectonic Map of Europe goes back to 1956, to the XXth Session of the International Geological Congress in Mexico. Geologists from many European countries enthusiastically undertook to work on the project proposed to the Congress by the late Professor A. A. Bogdanoff on behalf of the Soviet delegation. As a result, a manuscript map was demonstrated at the XXIth Session of the IGC in Copenhagen, and by the XXIIth Session of the Congress (New Delhi) the map was published. At that Session a resolution was passed to begin work immediately on the second edition of the map, abundant new information having been accumulated which necessitated modification of the legend.

The realization of the new project (up to 1971 under the indefatigable direction of Professor Bogdanoff with the active participation of many European geologists, in particular H.-R. von Gaertner, F. W. Dunning, M. Lemoine and others sponsored by M. J. Marcais, President of the Commission for the Geological Map of the World) took about ten years and only late in 1974 were the last sheets of the map given to the publishers. Thus we may hope that the colour proof of the map will be ready by the XXVth Session of the IGC in Sydney and in 1977 the map will come off the press.

19

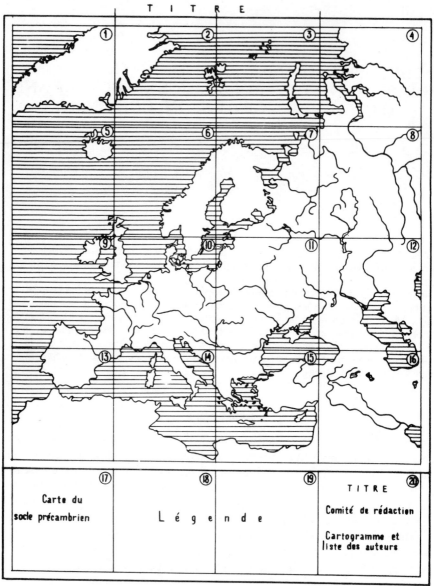

Fig. 1. Key to the International Tectonic Map of Europe (2nd edition)

The new map differs essentially from the first edition. First of all, Europe is shown in a broad frame which includes the Arctic shelf seas, the northern Atlantic, the whole of the Mediterranean together with the northern peripheries of Africa and Asia Minor, and a large part of the Transcaspian region (Turan) and West Siberia, including Taimyr and the Kara Sea (Fig. 1). Therefore, the new map consists of 16 sheets instead of 12 as in the first edition, the scale being the same

(1 : 2 500 000). We can note with much satisfaction that all the European countries, as well as some African (Morocco, Algeria, Tunisia, Egypt) and Asiatic (Turkey, Iran, etc.) countries took part in the collective work on the map.

Important changes have been introduced into the legend of the map though its basic principle—subdivision into regions according to the time of main deformation—is wholly preserved.

Within the East European craton and younger platforms, structures of the lower part of the sedimentary cover are shown and an attempt has been made to portray the internal structure of the basement, which was done more thoroughly in a special 1 : 10 000 000 insert.

Subdivision of Precambrian complexes of shields and massifs (Baltic, Ukrainian) has been much improved, mainly due to progress in radiometry. Drilling and geophysical data permit the basement structure of the White Russian and Voronezh massifs to be shown down to the zero contour-line. The same has been done for the Urals and Turgay trough and for the Palaeozoic massif of Central Kazakhstan and, as a result, the correlation of the structures of the Urals and Kazakhstan has become clear. Certain difficulties were involved in the portrayal of heterogeneous tectono-metamorphic reworking of ancient rocks, in particular the Archaean and Early Proterozoic.

Ancient complexes of Palaeozoic and Alpine folded areas are subdivided by the time of their main deformations and degree of reworking, whereas on the old map these ancient basement complexes were generalized and merely indicated as C_0, V_0, A_0. As regards the main geosynclinal complexes, it was decided to subdivide them by the main phases of folding, instead of structural stages, the latter being sometimes rather contrived (where there are no disconformities between structural stages). Such subdivision is undoubtedly more in line with the main principle of the map but in this case it is necessary to indicate the stratigraphic volume of rocks involved in a given folding episode and difficulties arise in showing a series of subsequent deformations.

Abundant new data on the structure of the submarine peripheries of the continent permit the legend for land areas to be applied offshore. Traditional representation of bathymetry of the deep waters of the Atlantic and Mediterranean is supplemented by several structural symbols, which is an advance on straightforward geomorphological portrayal.

Main Structural Subdivisions

As is generally known, half a century ago Stille subdivided Europe into areas of different time of consolidation: the Precambrian *Eo-Europa,* the Caledonian *Palaeo-Europa*, the Hercynian *Meso-Europa,* and the Alpine *Neo-Europa* (Stille, 1924; Fig. 2). This subdivision remains valid and the limits of the areas are essentially the same, but in Stille's scheme there is an area in between *Eo-,** *Palaeo-* and

* *Editors' note:* 'Ur-Europa' was the original usage of Stille (1924), 'Ur' being the Germanic prefix for 'very ancient'. Most authors since have preferred 'Eo-Europa' for conformity with the other Greek prefixes. The two terms are therefore synonymous.

Fig. 2. Stille's tectonic subdivisions of Europe

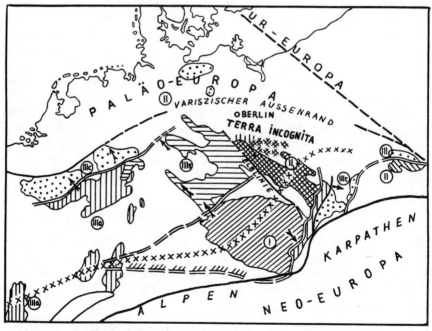

Fig. 3. Stille's 'Terra incognita' in northern Europe

Meso-Europa, the consolidation time of which was not clearly defined and which was called by Stille *Terra incognita* (Fig. 3). This is the area of the Polish–German lowland and the North Sea where the basement is very deep (a borehole 7 km deep north of Berlin reached only to the Namurian): von Bubnoff (1926) called it the Mid-European depression. The age of the basement in this depression is unknown but the following information is relevant (Fig. 4).

The fact that highly metamorphic rocks identical in composition and absolute age with those in the southwestern block of the Baltic shield were reached by drilling in the south Danish uplift (Ringköbing-Fyn high), as well as the platformal character of the Early Palaeozoic sediments in the north Danish downwarp, permit Denmark to be included, together with the adjacent parts of the North and Baltic Seas, in the East European craton, i.e. into *Eo-Europa,* as was suggested by Schatsky (1946). Thus, the southwestern border of the ancient craton follows the Teisseyre–Tornquist line only up to the southern coast of the Baltic, but from there it must run north of Rügen Island through the neck of Jutland.

The British Midlands may serve as additional evidence to make this point clear. Early Palaeozoic rocks here, as in Denmark, are positively of a platform nature; the basement is Late Precambrian, possibly of Late Riphean or Vendian age. But isotope ages indicate extensive Baikalian (Cadomian) reworking.

The third line of evidence is the Brabant massif. In its southern segment, the Caledonian folded complex is clearly identifiable, while the middle and northern parts of the massif comprise, according to the latest data, Late Precambrian rocks discordantly overlain by Cambrian. They are most likely the Baikalian complex.

Undoubted Baikalian complex—Riphean greenschists overlain by Vendian molasse—was reached in boreholes just to the southwest of the Swiety Krzyz Mountains and in the basement of the Carpathian foredeep in Poland. In the southeast it borders a zone of Caledonian folding and metamorphism which is bounded on the other side by an ancient block equivalent to the Bohemian massif. The Baikalian complex is traced from southwestern Poland along the Fore-Carpathian region as far as Dobrogea in Romania and the Black Sea.

Finally, in Rügen Island and Pomorze in Poland, in the southern Swiety Krzyz Mountains and beyond in the Soviet Fore-Carpathian region and Moldavia, a zone of thick Ordovician–Silurian slates overlain by Devonian red beds is traced, analogous to the southern periphery of the Brabant massif. Phyllites of uncertain age (either Early Palaeozoic or Late Riphean) were discovered by drilling in the extreme south of Jutland.

How can these rather contradictory data on the basement of the Mid-European depression, on the whole indicative of its heterogeneity, be co-ordinated? It seems that cratonization of the platform was predominantly Baikalian, which conforms to the intermediate position of this area between the ancient craton, the Caledonides and the Hercynides. Therefore, following Zhuravlev (1971) and Muratov (1975), we may consider the Epi-Baikalian platform to be generally similar to the Timan–Pechora platform on the other side of the craton. At the same time one should be aware of the fact that within this platform there may be both median massifs (beside the Midlands, there is the East Elbian massif, traced by geophysical data) and Caledonian intracratonic folded zones, such as the Pomorze–Moldavian zone, trending along the Teisseyre–Tornquist line.

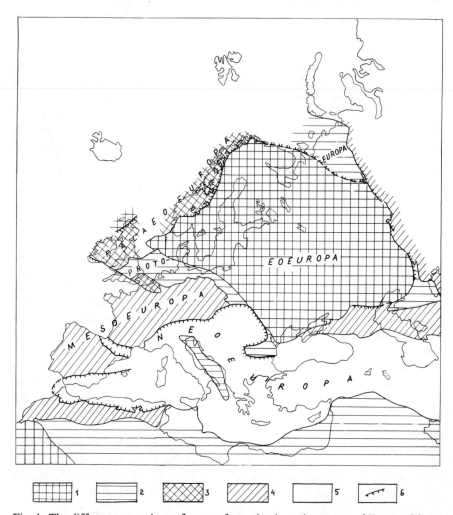

Fig. 4. The different generations of craton formation in various parts of Europe. (1) Ancient craton; (2) area of Epi-Baikalian cratonization; (3) area of Epi-Caledonian cratonization; (4) area of Epi-Hercynian cratonization; (5) Alpine folded belt; (6) front of Alpine and older nappes

Thus it appears from the above that not four, but five areas of cratonization of different generations occur in the European continent (Fig. 4): the Pre-Baikalian *Eo-Europa*, the Baikalian *Proto-Europa*, the Caledonian *Palaeo-Europa*, the Hercynian *Meso-Europa*, and the Alpine *Neo-Europa* (the cratonization of the latter is not yet finished).

Pre-Baikalian Blocks Beyond *Eo-Europa*

The East European craton was formed as a single entity between 1750 and 1650 Ma ago. The intrusion of anorogenic granites of the Rapakivi type along the

Fig. 5. Reworking of the East-European craton during the Gothian and Dalslandian episodes. (1) Ancient craton; (2) Archean massifs within craton, partly reworked; (3) Gothian and Dalslandian reworking; (4) Rapakivi type granites; (5) Dalslandian and older massifs within Baikalian and younger folded belts; (6) Late Precambrian basalts

western and southwestern peripheries of the craton and the beginning of formation of aulacogens within the Russian platform took place at the same time. Only the western portion of the craton protruding into southern Norway, SW Sweden, and, apparently, into Denmark, Poland, and part of White Russia and the

Ukraine* remained mobile after that time and was involved in Gothian and Dalslandian reactivation (Fig. 5).

Correlation between the internal structure of the craton and its boundary faults, especially in the southwest and east, proves that these boundaries are secondary and that originally the craton must have continued far beyond its present limits. This is corroborated by the fact that within the rest of *Eo-Europa* (as well as *Proto-*, and *Paleo-*, and *Meso-*, and *Neo-Europa*) there are blocks of high-metamorphic rocks (commonly amphibolite, rarely granulite-facies), mainly gneisses, evidently different from the greenschists and phyllites of the Baikalian (Cadomian) complex. Only in the Normandy massif (Domnonea of Pruvost) does this Pre-Baikalian complex (Pentevsian) yield ancient Archaean dates, while in other places there are dates of about 1000 Ma, i.e. Dalslandian (Grenvillian). In most cases these dates do not indicate the age of initial metamorphism: as with the Sveco–Norwegian block, the dates probably represent 'rejuvenation' of blocks of older, Early Precambrian continental crust.

But in this case it must be admitted, firstly, that the Baikalian and early geosynclinal systems in Europe originated on a broken basement like that of the East European craton, and, secondly, that the basement of western and central Europe was almost totally reactivated in the Dalslandian–Grenvillian epoch, that is, before the beginning of the Baikalian geosyncline—the first generation of the Mediterranean geosynclinal belt. Furthermore, the foregoing may imply that the East European and African cratons were originally a single entity. It is worth noting in passing that a major part of the African craton opposite western and central Europe was also reactivated in the Late Precambrian.

An interesting feature of the Pre-Baikalian blocks in Europe is late acid volcanicity, altered into porphyroids and even augen-gneisses; a typical example is the Ollo-de-Sapo series in the Iberian Meseta. These rocks resemble the Swedish sub-Jotnian, the Yetti–Eglab series of the Reguibat massif in Africa, and others.

Rocks of the sedimentary cover proper are not preserved on the European ancient blocks: they appear only in northern Africa (the Kabylian massifs in Algeria) and in Syria and Iran, that is, in the southernmost zone of the massifs bordering the African–Arabian platform.

Distribution of Oceanic Crust in Late Proterozoic and Palaeozoic Geosynclines of Europe

From the foregoing it might follow that all the Late Precambrian–Phanerozoic geosynclines in Europe were formed on the ancient continental crust and were both intracratonic and epicontinental. This is undoubtedly true of the major areas of these systems, and is corroborated by the rock composition of the geosynclinal complex. But in general such a conclusion is probably erroneous because ophiolite

* The Ovruch series in the northern Ukrainian shield and its presumed continuation in the zone of the Dnieper–Donets trough.

formations, which for good reasons are now considered to be relics of oceanic crust, occur in Baikalian, Cadomian and Hercynian geosynclines of Europe and of adjacent areas.

It is known that these rocks appear in the British Caledonides on both sides of the Scottish Midland Valley, as well as in the Scandinavian Caledonides at the base of the Lower Palaeozoic of the innermost nappes. Therefore many geologists agree that the central segment of the Caledonian geosynclinal system originated on newly formed oceanic crust. But it remains uncertain whether it was a veritable ocean (the northern Protoatlantic) or, rather, resembled the present Mediterranean.

It is probable that the Caledonian folded zones in southeast England and Brabant, as well as those along the southwestern edge of *Eo-Europa*, are epicontinental structures. The same is apparently true of the Baikalian basement of *Proto-Europa*.

The problem of *Meso-Europa* is essentially different. It is interpreted on the one hand from the classical non-mobilist point of view, assuming it to have originated entirely on the ancient continental crust (Zwart, 1967; Krebs and Wachendorf, 1973), and on the other hand from the plate tectonics point of view (Burrett, 1972; Laurent, 1972; Riding, 1974; etc.). The problem is complicated by the fact that most of the original *Proto-Europa* became a part of *Meso-Europa* at the beginning of the Palaeozoic, and *Meso-Europa* was incorporated into *Neo-Europa* at the beginning of the Mesozoic. In the course of these processes they were subjected to complex movements, most likely including large-scale horizontal and rotational movements. Therefore, to analyse the development of the Baikalian and Hercynian geosynclines in central and southern Europe, it is necessary to analyse structures that had been formed by the beginning of both the Palaeozoic and Mesozoic.

It is by no means the object of this report to make such palaeotectonic reconstructions or to suggest new palaeodynamic models. But I should like to draw attention to some characteristic features of the Pre-Alpine structure in the part of Europe under consideration which suggest, on the one hand, the origin of Late Precambrian–Palaeozoic geosynclines (not only on the continental, but also on the oceanic crust) and, on the other hand, the very complex structure of this area. First of all we shall deal with relic deep-sea trenches with oceanic crust, as evidenced by ophiolite belts (Fig. 6).

The northernmost ophiolite belt of *Meso-Europa* is correlated with the southern edge of the Cornwall–Devon massif. Here, on the Lizard Peninsula, an apparently Precambrian series of crystalline schists, amphibolites, gabbro and serpentinites is exposed, thrust against the Devonian. As Cornwall and Devon are the continuation of the Mid-European Rheno–Hercynian zone and to the south, on the French coast, there is the Normandy massif (Domnonea)—the western continuation of the Mid-German crystalline sill (Mitteldeutsche Schwelle)—the ophiolite belt may be traced here into the continent. Ophiolites are not known here, but stronger metamorphism of Early and Middle Palaeozoic rocks is noticeable.

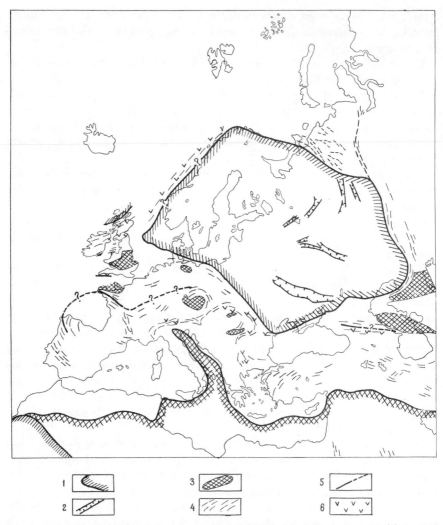

Fig. 6. Relation of ophiolite belts to cratons and folded complexes in Europe. (1) Ancient craton; (2) aulacogens (rifts) within craton; (3) reworked ancient massifs; (4) Baikalian folded complexes; (5) ophiolite belts; (6) spilitic rocks

One of the most distinct ophiolite belts of *Meso-Europa* trends between the central and southern parts of the Armorican massif, from Baie d'Audierne in the west (Cogné, 1974). A rather typical, though badly disturbed, cross-section is observed here: ultrabasites, amphibolites, gabbroids, spilites. Its correlation with the Brioverian shows that originally this series most likely underlay the Brioverian, which indicates that the latter (Late Riphean according to the Soviet geochronological scale) originated on oceanic crust.

At present, it is almost universally accepted that before the Bay of Biscay was formed in the Mesozoic, Armorica and the Iberian Meseta were a single entity.

Indeed, in the northwest Meseta there is a similar ophiolite zone to that of Cabo–Ortegal–Braganca, whose rocks must also be considered as originally underlying the Baikalian (Cadomian) geosynclinal complex.

This ophiolite belt bends northwest, forming an arc that must have girdled the Iberian–Aquitanian block of the continental crust, including northern Spain and southwestern France. The block was cut first by the Hercynian and, later, by the Alpine intracratonic Pyrenean geosyncline. Unfortunately we do not know for certain the continuations of the arc. Its southwestern end must have been in Morocco, and the eastern extension must have flanked on the north the Central French and Bohemian–Wisla massifs. Small ophiolite outcrops, probably also of Precambrian age, occur in the Erzgebirge, near the junction of the Bohemian massif with the Saxo–Thuringian zone, and in the Sudetic Mountains. Algonkian spilite volcanism is the echo of the same process of formation of oceanic crust during the origin of the Mediterranean belt in the Late Precambrian. The Saxonian ophiolite belt is most probably the eastern continuation of the Armorican one; it trends under the Carpathians and, probably, into Dobrogea, reappearing in the Northern Caucasus.

In recent years, new data on the Northern Caucasus have been obtained. The structure of the Front Range with its abundant migmatites was believed to be rather simple, and ultrabasic rocks occurring here were supposed to be of intrusive origin and Carboniferous age (C_1/C_2). It is now certain that large-scale nappes are developed here, the overthrust slices containing ultrabasites, gabbro and greenschist volcanics. The most probable direction of overthrusting is northerly, towards the Palaeozoic miogeosyncline lying in the Fore-Caucasus region. 'Roots' of nappes lie, apparently, in the fault zone between the zones of the Front and Main Ranges. The interpretation of the nature of the latter has been changed; it was formerly considered a geanticline inherited from the beginning of the Palaeozoic, while now it is assigned to the Hercynian metamorphides. Therefore, the following zones may be traced in a cross-section of the Hercynian system in the Northern Caucasus: the miogeosyncline of the Fore-Caucasus (Saalic folding); the eugeosyncline of the Front Range with nappes (Sudetic phase); the metamorphic zone of the Main Range with granite-gneissic domes (Bretonian? phase). Further south there is a trough-zone with a complete Palaeozoic marine sequence practically devoid of volcanics, which after the Permian and before the Jurassic was metamorphosed to phyllite grade. The relative position of the first three zones closely resembles the Mid-European Hercynides, and that of all four, the Kun-Lun system, with which the Caucasus was probably connected via Turkmenia, NE Iran and Afghanistan. In the west, the last zone passes into the Crimea Mountains and disappears but a very similar formation is observed in the Bukk Mountains south of the Western Carpathians, connected in their turn with the Southern Alps and Dinarides.

Ultrabasic rocks in the Northern Caucasus are at least of Pre-Silurian age, dated by the 450 Ma granites which penetrate them.

In the Balkans, both in Stara Planina and in the Serbian–Macedonian massif, there is a very typical diabase-phyllite formation. It is overlain transgressively by the Ordovician, and its Early Cambrian age has been proved by several fossil discoveries in the Serbian–Macedonian massif. It is possible that the lower part of the formation is of Vendian age. Equivalents of this formation are known along virtually the whole length of the Carpathians (the Corbu and Tulgesh series in Romania, the Delovets series in the Ukraine, Rakoviecz-Jelnica in the Gemerian zone, Slovakia). In the Iron Gate region (the Danube) the basic-ultrabasic massif is likely to be connected with that formation. In the Southern Alps (the Carnic Alps, Karawanken) the Ordovician is also discordantly underlain by a greenschist psammitic-pelitic series comprising thick porphyroids with albite. These may be relics of a single trough (double in the Carpathians) with spilite-keratophyre and,

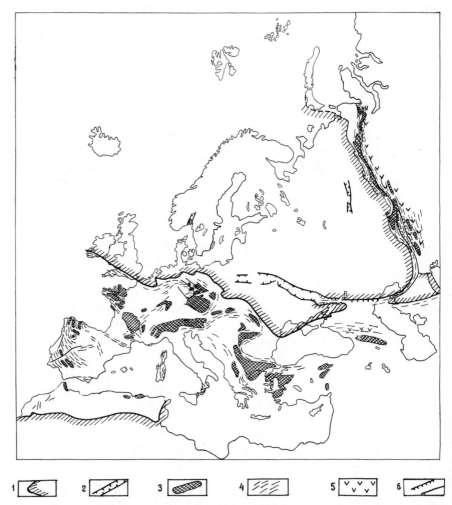

Fig. 7. The major Hercynian structures of Europe. (1) Pre-Hercynian platform area; (2) Hercynian aulacogens (rifts); (3) Pre-Hercynian reworked massifs; (4) Hercynian folded complexes; (5) basic volcanics; (6) thrusts and major faults

in places, ultrabasic magmatism. Its northern branch trends in the east to the Black Sea: on the opposite coast in Georgia there is a similar metamorphic formation in the Dzirula massif of Vendian–Early Cambrian age corroborated by palaeontological data.

It is curious that the Late Riphean–Vendian–Early Cambrian was the main period of formation of oceanic crust in the central Asiatic region (Sonnenschain, 1973) and, probably, in the whole of the Uralian–Mongolian belt.

The central and south European equivalent of this period is the initial Cadomian, or Baikalian, stage of development of the Mid-European geosyncline which ended in the Sardic tectogenetic phase of the Late Cambrian, while the Hercynian geosynclinal cycle proper began in the Ordovician. Prior to the Sardic, there were typical geosynclinal troughs with oceanic crust separated by microcontinents: (Normandian, Franco–Bohemian, Aquitanian–Iberian, Pannonian, Rhodopean) but Sardic folding, metamorphism and granitization resulted in the formation of quasi-continental crust. Therefore the Hercynian geosynclinal process developed in Europe mainly in episialic conditions, which is most strikingly evident from the absolute predominance of granitoid over basic-ultrabasic magmatism (Zwart, 1967).

Nevertheless, the existence is not excluded of troughs with thinned crust, filled in with slate-diabase rocks (the Rhenohercynian trough) or Benioff-type subduction zones (see a recent work by Anderson (1975) on the Harz Mountains).

The poor development of nappes in the European Hercynides, unlike the Caledonides and Alpides, may be connected with the development of the Hercynian geosynclines in episialic conditions and the absence of any wide oceanic '*hiatus*'. The nappes are known mainly along the periphery of the folded systems, at the boundary with areas of earlier consolidation. These are the nappes along the southern edge of the 'Coal Channel' in Western Europe, of the Harz, in the western and southern peripheries of the Asturian coal basin and, apparently, of the Montagne Noire region of southern France. Among nappes of the interior zone is the Champtoceaux nappe connected with the ophiolite zone of southern Armorica.

Origin of Alpine Geosynclines in *Neo-Europa*

The main problem of the tectonics of *Neo-Europa* is that of the original distribution and extent of areas of oceanic crust, as well as their time of origin and their connection with older, especially Hercynian, structures. In the decade following publication of the first edition of the Tectonic Map of Europe and its accompanying explanatory note, it has become evident that the Alpine geosynclinal belt originated entirely within the Hercynian belt only in the west while from the southwestern Dinarides and Hellenides, Anatolia, Transcaucasus and farther southeast a considerable part of the belt was formed on the Epi-Baikalian platformal basement. The cover of this Epi-Baikalian platform being very much like that of the African–Arabian craton, it may be called Peri-Gondwanan. The general shift of the Alpine belt south relative to the Hercynian one is accompanied here by a similar shift of zones of downwarping, folding and magmatism. At the junction of the Epi-Hercynian (Peri-European) and Epi-Baikalian (Peri-Gondwanan) platforms there was a zone of deep-sea troughs of continuous marine sedimentation in the Mid and Late Palaeozoic, as well as of folding and epizonal metamorphism, locally associated with the formation of granites in the Triassic. This zone was dealt with above. Here it should be pointed out that it is displaced between its western segments (the Carnic Alps, Bukk Mountains, Dinarides–Hellenides) and eastern segments (the Crimea Mountains, southern slope of the Greater Caucasus and further east), a fact which is not yet explained. South of this zone the Hercynian reactivation (metamorphism, magmatism) gradually dies out.

The available data suggest that the Hercynian development at the end of the Palaeozoic and beginning of the Triassic resulted in the restoration of the continuity of the continental crust between Europe and Arabia–Africa throughout the areas represented on the new Tectonic Map (Argyriadis, 1975; Khain, 1975). Accordingly, the Permian and Early Triassic basins of southeastern Europe, northeastern Africa and Asia Minor should be regarded as a vast epicontinental sea but not as a true ocean. Oceanic crust of this age appears only in the Tibetan Himalaya (recent observations by Chinese geologists). True, the presence of Palaeozoic ophiolites in the Dinarides assumed by some Yugoslavian geologists remains a point of much controversy. If this hypothesis were to be corroborated it would much affect the concepts stated above and would suggest a certain succession of *Palaeo-Tethys* to *Tethys* proper and, it must be admitted, would facilitate various mobilist reconstructions. For the time being the idea of the absence of a direct structural heredity between Palaeozoic and Mesozoic structural plans of the Mediterranean, which has been vividly demonstrated by Argyriadis, seems more convincing.

The formation of new oceanic crust in the *Tethys* area (Mesogea) began in the Middle-Late Triassic and progressed, apparently, until Late Jurassic, followed by the reverse process of 'clustering' (Peive, 1969). Relics of this crust can be observed as strongly reduced ophiolite belts in the Western Mediterranean (lherzolites of the Pyrenees, ophiolites of the Betic Cordillera, Rif, Tell), while from the Alps they trend in a wide zone of complex structure towards Asia Minor.

A particularly difficult problem in the Alpine Mediterranean is to determine the character of the area of original oceanic crust. Was this area a vast oceanic basin a thousand or more kilometres wide, or a complicated system of troughs first hundreds of kilometres wide, separated by wider blocks of continental crust, that is, microcontinents? In other words, what was *Tethys*? Was it an ocean of Atlantic type or similar to the present Mediterranean Sea but more open?

It is quite evident that in the present structural plan there are uplifts of ancient Precambrian continental crust dividing ophiolite belts, and a great number of the latter, especially beginning at the Hellenides and extending further southeast and east (Fig. 8). The question is how these uplifts should be interpreted. One of the hypotheses is that they are nappes overthrust as a result of northward movement of Gondwana upon *Tethys* (Peive, 1969). Separate massifs do in fact exhibit traits

Fig. 8. The major Alpine structures of Europe. (1) Platform area; (2) rifts within platform; (3) Alpine nappes front and main faults; (4) Alpine folded complexes; (5) massifs of Pre-Alpine continental core; (6) ophiolite belts

of being overthrust upon the adjacent ophiolite zones, such as the massifs of Rhodope, the western edge of the Pelagonian, and the northern edge of the Arzakanian in the Transcaucasus. But it seems that these are comparatively restricted phenomena and they do not permit us to consider these massifs as being entirely allochthonous. The latter is scarcely probable in Iran and Afghanistan where the distance of the necessary overthrusting exceeds all conceivable figures.

Another hypothesis pertains to the southwestern zone of massifs, including Pelagonian and Menderes (Aubouin, 1973; Ricou *et al.*, 1974, etc.). According to this hypothesis, all the massifs are regarded as a part of the northern periphery of the Arabian–African continent, exposed in vast tectonic inliers. Correspondingly, the Subpelagonian–Taurian ophiolite belt nearest to this continent is considered entirely allochthonous, the 'motherland' of its ophiolites supposedly lying to the north of the belt of median massifs. Thus, it may be assumed that within *Tethys* there was a single, but very wide, zone of oceanic crust.

In favour of this hypothesis are both structural data and the consideration that it is unnatural for the southern ophiolite belt to lie on the very edge of the platform. It is remarkable that Alpine metamorphism is most intense in massifs of this type; in fact, they may be huge granite-gneissic domes obducted in the rear of the overthrust ophiolite masses due to density inversion. An analogous explanation is suggested for similar domes in the eastern side of the Urals (A. S. Perfilief) but in both cases the source of high heat flow which promoted remobilization of the material of the granitic-gneiss layer of the continental crust should be indicated. Such remobilization would be natural above a Benioff zone plunging under the continent and in this case the obduction/subduction relations would be like those suggested by Coleman (1971).

Dealing with this hypothesis, one should explain why the ophiolite nappes are concentrated within certain belts. This may be due to a deep trough uncompensated by sedimentation and bounded on the rear by a fault trending along the periphery of the platform. Ophiolite material, in the form of overthrust sheets, melange and olistostromes, was deposited in just such troughs and in addition squeezed by the rear overthrust. This mechanism may be invoked to explain the tectonic structure of the westernmost ophiolite zone of the Urals—the Sakmar zone (with its northern continuation). Deep-sea siliceous sediments and some basic volcanics may be autochthonous here, the allochthonous rocks being only serpentinites, gabbro and the rest of the basic volcanics.

But however tempting the second hypothesis is, it is faced like the first one, with the 'problem of space' as the Alpine belt trends farther east, where the distance between the northern- and southernmost ophiolite belts exceeds 1000 km. Nevertheless, in some places the hypothesis may be valid again, for example in the Afghan–Pakistan region for the elucidation of the relations between the Gilmend continental block and ophiolite belts east and southeast of it (Karapetov *et al.*, 1975).

The available data are obviously insufficient to treat every ophiolite zone, automatically and unequivocally, as the outcome of an independent oceanic trough and to consider its ophiolite nappes as formed *in situ* as a result of its closing. Neither is it reasonable to consider as relatively autochthonous only one ophiolite zone in each section of the Alpine belt (and of other belts) and, correspondingly, to consider all blocks of the continental crust between ophiolite zones as either overthrust nappes or tectonic inliers of the adjacent continental plates. Certainly, it cannot be ignored that at least some of these blocks, especially the largest of them, are microcontinents (median massifs) inside a geosynclinal belt (paleoocean).

In the northern peripheral part of the Alpine belt, including the Alps and Carpathians, a trough-microcontinental structure for the belt is more probable than an oceanic structure proper. In such structural conditions 'freedom of movement' is especially important, in particular the rotation of microcontinental blocks

(microplates) which, taking into account their great proneness to internal deformation, makes a strictly geometric theory of plate tectonics inapplicable.

Only detailed field research, drilling and geophysical methods will permit us to
choose one of the possible alternatives in each case.

Rift Systems, Old and Young

In the works of Stille the rift zone of the Mjösa Lake–Mediterranean Sea was
already recognized including, in particular, the Rhine Graben. The zone was
developing from Permian to Recent time. But for years this rift system was considered a unique phenomenon and even when, on the basis of drilling data,
Schatsky (1946) formed an idea of Riphean and Devonian aulacogens* on the
Russian platform, the analogy between these two groups of structures was for
some time overlooked. Thereupon buried rifts were discovered in the North Sea
basin and on its southern prolongation in the Netherlands and German Federal
Republic. It is beyond question now that as separate parts of the European continent were cratonized, they became areas of rift formation which, thus, took place
during all the tectogenetic stages, beginning with the Proterozoic. It should be
pointed out that in Europe, along with fully developed rifts, there are many
embryonic and abortively developed ones, for the most part buried.

The earliest generation of rift structures (in a broad sense) may be considered
the Mid-Proterozoic Pechenga and Imandra–Varzuga fault-bounded downwarps
of the Baltic Shield in the southern Kola Peninsula. These 'proto-aulacogens' are
filled in with clastic red beds and basic volcanics of considerable thickness.

The next very important generation are the Riphean aulacogens of the Russian
platform (Fig. 9) forming a complicated network resulting from the fracturing of
the ancient craton soon after it was developed. They predetermined the location
of Phanerozoic syneclises: for example, the largest of them—the Moscow syneclise—originated at the triple junction of the Mid-Russian, Moscow and
Pachelma aulacogens.

The formation of paleorifts-aulacogens resumed on the Russian platform in the
Middle Devonian–Early Carboniferous, that is, at the beginning of the Hercynian
cycle. The largest of these is the Pripiat–Donets aulacogen which is superimposed, as recently ascertained by deep seismic sounding, on an older (Riphean)
and narrower aulacogen trending under the Donets basin. Such Devonian revival
of Riphean aulacogens is characteristic not only of the Pripiat–Donets aulacogen;
it was also established in the Kazan–Sergiev aulacogen in the east of the Russian
platform. The echoes of the same Devonian rift formation are observed in the
north of the Baltic shield in the central part of Kola Peninsula, where alkaline intrusions occur. The graben-trough of the Midland Valley in Scotland, exhibiting
alkaline-basic volcanism, is a further example of a Devonian–Carboniferous rift
structure.

---

* Aulacogens are long-lived, deeply subsiding troughs, at times fault-bounded, that extend at high angles from geosynclines far into adjacent foreland platforms (Hoffman,
Dewey and Burke, 1974).

Fig. 9. The major rift systems of Europe. (1) Late Precambrian rifts; (2) Mid-Palaeozoic (mainly Devonian) rifts; (3) Mesozoic (mainly Jurassic) rifts; (4) Cenozoic rifts

The formation of rifts was resumed at the end of the Hercynian stage, in the Permian, the main area of tectonic activity being the south of the Baltic shield, with the conspicuous Oslo Graben, and in the adjacent North Sea region. In the Permian, the Danish–Polish aulacogen originated along the southwestern border of the East-European craton, as well as the Celtiberian aulacogen within the Iberian Meseta. At the base of the Anglo–Paris basin there are Permian rifts trending northeast and northwest. The Manych zone of Late Permian–Early Triassic graben-troughs trends from the Don to the northern Caspian.

    The next epoch of rift formation accompanied and succeeded the Early Cimmerian orogeny; it is dated as Rhaetic–Liassic–Dogger. At that time, numerous grabens were formed on the eastern side of the Urals and in the Trans-Uralian region, as well as in the Fore–Black-Sea zone of graben-troughs along the junction of the East European craton with the younger Scythian platform from Moldavia to the eastern coastal region of the Azov Sea. The formation of rifts resumed in the North Sea region and in its southern periphery in the Netherlands, GFR and further west in southern England (Whittaker, 1975).

The phase of rifting connected with the Late Cimmerian tectogenetic epoch was relatively weak within the European continent proper. At that time the main events occurred in the Atlantic, in particular in the Bay of Biscay and English Channel. Tentatively, the formation of rifts leading to the development of the Black Sea depression could be attributed to this phase.

Rifting within the continent and around it intensified abruptly in the Oligocene–Miocene after the Pyrenean and Savic contractional paroxysms, and the Rhine–Saone–Rhône and, apparently, Algeria–Provence grabens were formed. This system extends northwards to the North Sea and Norwegian Sea and southwards into Africa. The eastern branch of the East African rift system extends northwards into Anatolia,* while its more westerly branches contributed to the formation of the depressions of the Aegean, Ionian and Adriatic Seas. Side by side with longitudinal (the Atlantic trend) rift zones, latitudinal zones (the Mediterranean trend) developed at that time. These are the Alboran zone, the zone along the Main North Anatolia fault (from the north of the Aegean Sea to the Sevan Lake in Armenia), and the zone of the Bay of Corinth–Menderes.

By analogy with more reliable examples, it may be presumed that the formation of the Skagerrak and Kattegat, the Gulf of Bothnia, the central Baltic Sea (apparently a very old rift occurs here), and the zone of subsidence from the Gulf of Finland through Lakes Ladoga and Onega to the White Sea, is connected with incipient rifting. It is probable that the latitudinal Barents trough between Bear Island and Scandinavia, the depressions in the Barents Sea floor (such as the St. Anne trough) and the channels separating Britain and Ireland, are also of rift origin.

There are, apparently, certain regularities in the distribution of rifts and their systems in time and space. Firstly, rifting phases correspond to periods of high tectogenetic activity in general, for example to periods of folding. It should be pointed out that they either accompany the latter and are developed in the rear of zones of compression and formation of folds and overthrusts (for example the Uralian foredeep and the eastern slope of the Urals and Transuralian region), or immediately follow the folding and overthrusting, as superimposed structures. Secondly, in areas of earlier cratonization, rift zones originate parallel to geosynclines and newly formed oceanic areas; hence the Atlantic and Mediterranean trends of young rifts in Europe and its surrounding seas, which were recently recognized in the North Sea and Atlantic by Whiteman *et al.* (1975).

Thirdly, as rift zones and particularly their junctions (triple junctions) develop, they degenerate either into shelf seas, syneclises, or depressions of inland seas of

* Its weakened continuation can be traced throughout the Black Sea, Western and Fore–Caucasus, up to the southeastern Russian Platform.

suboceanic type. Both in the first and especially in the second case the continental crust thins out and is replaced, in some way or other, by oceanic crust. In fact all areas of thinned, reduced continental and suboceanic crust within Europe and its surrounding seas are former (in part, present) areas of intense rifting. Many important topographic features of Europe and adjacent seas were also predetermined by rifts.

Fourthly, spasmodic parallel migration of rift systems in time is noticeable from areas of earlier to those of later cratonization. An example of such migration may be the following sequence of sublatitudinal rifts (from north to south): (i) the Moscow rift (Riphean); (ii) the Dnieper–Donets rift (Riphean?–Devonian); (iii) the rift north of the Black Sea (Jurassic–Early Cretaceous); (iv) a hypothetical Central Black Sea rift (Late Jurassic–Early Paleogene??); (v) the rift in the north part of the Aegean Sea—the Dardanelles—the Sea of Marmora—Northern Anatolia—Sevan Lake (Late Neogene–Quaternary).

Conclusions

In this paper I have dealt with some problems in analysing the abundant data synthesized in the Second Edition of the International Tectonic Map of Europe. Some considerations have been expressed on the problems touched upon, but the author does not claim to have solved them. The fact is that the Tectonic Map is only a factual base for development of such problems and it is its independence from any tectonic interpretations that makes it really valuable. Much work needs to be done to generalize the available information and collect additional data. An important part of this work will be the compilation of a series of palaeotectonic maps both on a Recent and on a palinspastic base. This task, like the compilation of the Tectonic Map of Europe, can be successfully coped with on the basis of international co-operation. The work on the Tectonic Map of Europe over a period of nearly twenty years is an excellent example of such collaboration, and it would not be sensible for the international association formed in the course of this work to bring its activities to a halt after the map has been created. Therefore let us take the necessary steps to continue our work in order to discover the main stages in the formation of the continent of Europe and, subsequently, to model the dynamics of this process.

In the meantime, geologists from all the continents will proceed with the work on the Tectonic Map of the World, an invaluable start on which has been made by the International Tectonic Map of Europe.

References

Anderson, T. A. (1975). 'Carboniferous subduction complex in the Harz mountains Germany', *Bull. Geol. Soc. Amer.*, **86**, 77–82

Anthonioz, P. M. (1975). 'Quelques réflexions sur la géologie des roches basiques et ultrabasiques du Cap Lizard (Cornwall)', *C.R. Acad. Sc. Paris*, **280**, D-399

Anthonioz, P. M. and Correa, A. V. (1973). 'Essai d'interprétation des associations basiques et ultrabasiques polymétamorphiques précambriennes du Nord-Ouest de la Péninsule Iberiqué, *C.R. Acad. Sc. Paris*, **277**, D-1105

Argyriadis, I. (1975). 'Mesogee permienne, chaine hercynienne et cassure tethysienne', *Bull. Soc. géol. France*, **7**, t. XVII, 56–70

Aubouin, J. (1973). 'Des tectoniques superposées et de leur signification par rapport aux modèles géophysiques: l'exemple des Dinarides; paléotectonique, tectonique, tarditectonique, néotectonique', *Bull. Soc. géol. France*, **7**, t. XV, 426–60

Bubnoff, S. V. von (1926). *Geologie von Europa*, Bd. 1, B

Burrett, C. F. (1972). 'Plate tectonics and the Hercynian orogeny', *Nature*, **239**, 155–7

Cogné, J. (1974). 'Le massif Armoricain', In *Geologic de la France*, **1**, P., Doin

Coleman, R. (1971). 'Plate tectonics emplacement of upper mantle peridotites along continental edges', *J. Geophys. Res.*, **76**, 1212–22

Floyd, P. A. (1972). 'Geochemistry, origin and tectonic environment of the basic and acidic rocks of Cornubia, England', *Proc. Geol. Assoc.*, **83**, 385–404

Hoffman, P., Dewey, J. F. and Burke, K. (1974). 'Aulacogens and their genetic relation to geosynclines, with a Proterozoic example from Great Slave Lake, Canada', in Dott, R. H. and Shaver, R. H. (eds.), *Modern and ancient geosynclinal sedimentation, Soc. Econ. Pal. Mineral. Spec. Publ.*, **19**, 38–55

Ivanov, S. N., Perfiliev, A. S., Efimov, A. A., Smirnov, G. A., Necheuvnin, V. M. and Ferstatter, G. B. (1975). 'Fundamental features in the structure and evolution of the Urals', *Amer. J. Sci.*, **275A**, 107–30

Karapetov, S. S., Sonin, I. I. and Khain, V. E. (1975). 'On some important peculiarities of the structure and development of the Afghan–Pamirs segment of the Alpine fold belt of Eurasia', *Moscow State University Bull.* (*Vestnik*) (Geol), **3**, 38–46

Khain, V. E. (1975). 'Structure and main stages in the tectono-magmatic evolution of the Caucasus: an attempt at geodynamic interpretation', *Amer. J. Sci.*, **275A**, 131–56

Krebs, W. and Wachendorf, H. (1973). 'Proterozoic–Paleozoic geosynclinal and orogenic evolution of Central Europe', *Bull. Geol. Soc. Amer.*, **84**, 2611

Laurent, R. (1972). 'The Hercynides of Southern Europe: a model', *Int. Geol. Congr.*, 24 sess., 363–70

Muratov, M. V. (1975). 'Middle European platform and its relation to East-European platform', *Bull. Soc. Nat. Moscow* L(3), **129** (In Russian)

Nicolas, A. (1972). 'Was the Hercynian orogen of Europe of the Andean type?' *Nature*, **236**, 221

Peive, A. V. (1969). 'Oceanic crust of the geological past', *Geotectonics*, **4**, 5 (In Russian)

Peive, A. V., Shtreis, N. A., Knipper, A. L., Bogdanov, N. A., Perfiliev, A. S. and Ruzhentsev, S. V. (1971). 'Oceans and the geosynclinal process', *USSR Acad. Sci. Rep.* (*Doklady*), **196**, 657–9

Peucat, J. J. and Cogné, J. (1974). 'Les schistes de la Baie d'Audierne (Sud-Finistère): un jalon intermédiaire dans le socle antécambrien entre la Meseta ibérique et les régions sud-armoricaines', *C.R. Acad. Sci. Paris*, **278**, D-1809

Ricou, L. E., Argyriadis, I. and Lefevre, R. (1974). 'Proposition d'une origine interne pour les nappes d'Antalga (Taurides occidentales)', *Bull. Soc. géol. France,* XVI, 107–11

Riding, R. (1974). 'Model of the Hercynian foldbelt', *Earth Planet. Sci. Let.*, **24**, 125

Schraeder, E. (1973). 'Probleme tektonischer Untersuchungen im Orogen, speziell in den Varisziden. Veröff. Zentralinst', *Physik. d. Erde*, N14, T. 2, 273–302

Schatsky, N. S. (1964). 'On troughs of the Donetz type', selected works, II, M. (In Russian)

Schatsky, N. S. (1946). 'Main features of the structure and development of the East European platform', *Bull. Izvestia USSR Acad. Sci.* (Geol), N1. (In Russian)

Sonnenschain, L. P. (1973). 'The evolution of Central Asiatic geosynclines through sea-floor spreading', *Tectonophys*, **19**, 213

Stille, H. (1924). *Grundfragen der vergleichenden Tektonik*, Borntraeger, Berlin, 443 pp.

Stille, H. (1951). 'Das mitteleuropäische variszische Grundgebirge im Bilde des gesamteuropäischen', *Beih. geol. Jb.*, **2**, 138 S

Whiteman, A., Naylor, D., Pergrum, R. and Rees, G. (1975). 'North Sea troughs and plate tectonics', *Tectonophys,* **26**, 39

Whittaker, A. (1975). 'A postulated post-Hercynian rift valley system in southern Britain', *Geol. Mag.,* **112**, 137

Zhuravlev, V. S. (1972). 'Comparative tectonics of the Pechora, Pericaspian and North Sea exogonal basins', M., 'Nauka'. (In Russian)

Zwart, H. J. (1967). 'The duality of orogenic belts', *Geol. Mijnb.,* **46**, 283

THE EARTH'S DEEP MANTLE AND CORE

J. A. JACOBS

Department of Geodesy and Geophysics, University of Cambridge, Madingley Road, Cambridge, UK

Abstract

Jacobs, J. A. (1976). 'The Earth's deep mantle and core', in Ager, D. V. and Brooks, M. (eds.), *Europe from Crust to Core*, Wiley, London.

The global distribution of earthquake zones and seismic recording stations is such that it is not possible to say much about conditions in the deep mantle and core beneath Europe. Lateral inhomogeneities in the Earth's deep interior do, however, exist and may be suspected beneath Europe. The evidence for regional variation of deep structure is reviewed.

Sharp increases in the frequency of geomagnetic field reversals at 85 Ma and 45 Ma B.P. are shown to correlate with major tectonic changes and both phenomena may be controlled by irregular features at the mantle–core boundary (MCB). It is suggested that such features may result from parts of the lower mantle occasionally becoming soluble in the outer core, or material from the core diffusing into the mantle. Such MCB topography may have a direct influence on core motions and, hence, on the frequency of geomagnetic field reversals.

Introduction

No direct measurements can be made of any of the physical properties of the Earth's deep interior. A fundamental problem in geophysics is the determination of some physical parameter (for example density) from a set of observations made at the surface of the Earth. This 'inverse' problem as it is called has received much attention during the last few years, particularly in the USA and the USSR.

The major source of information about the Earth's interior comes from seismology. At an epicentral distance of about 100°, rays from the source to the receiver intersect the mantle–core boundary (MCB). One method of obtaining information about conditions in the deep mantle and core beneath any region would be to study seismic records observed about 50° from that region received from an earthquake which occurred about 50° away in the opposite azimuthal direction. Fig. 1 is an azimuthal equidistant projection of the Pacific centred on the Large Aperture Seismic Array (LASA) in Montana. The seismic belt around the southwest Pacific from Kermadec to New Guinea is shown shaded. It can be seen that events from this seismic area, recorded at LASA, can give some information about the MCB in the general area below Hawaii. Fig. 2 is an azimuthal equidistant projection centred approximately on Lucerne, Switzerland. The lack of any seismic stations opposite any active earthquake zones shows that it is not possible

41

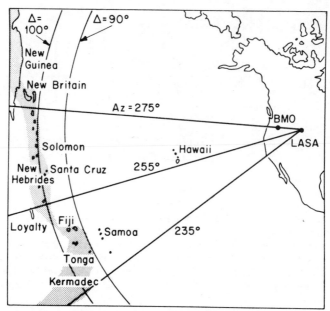

Fig. 1. Azimuthal, equidistant projection of the Pacific centred on
LASA. The seismic belt around the southwest Pacific is shaded
(after Needham and Davies, 1973)

to say much about conditions in the deep mantle and core beneath Europe. However, data from other areas in the world, to be discussed below, indicate that there are lateral inhomogeneities in the physical properties of the Earth's deep interior, and there is no reason to believe that some anomalies do not exist beneath Europe.

Lateral Heterogeneity in the Deep Mantle

Observations over many years of short-period seismic P waves have shown that the amplitudes observed from the same event vary remarkably little in the distance range 30–85°. Beyond 85°, however, the amplitude decreases steadily and is down by an order of magnitude at 100°. At some distance in the vicinity of 100°, rays from source to receiver intersect the MCB, and since the velocity of P waves is less in the core than at the base of the mantle, there is a shadow beyond this distance.

In 1969 observations were made by Phinney and Alexander of amplitudes at source–receiver distances which were well into the range in which the core should have shadowed out P waves. They found that amplitudes at a given distance were highly variable and depended on which region of the MCB was sampled. Since substantial lateral variations within the core are extremely unlikely, it appeared either that there is appreciable topography ('bumps') on the MCB or that the physical properties of the rocks at the bottom of the mantle are laterally variable—by perhaps as much as five per cent.

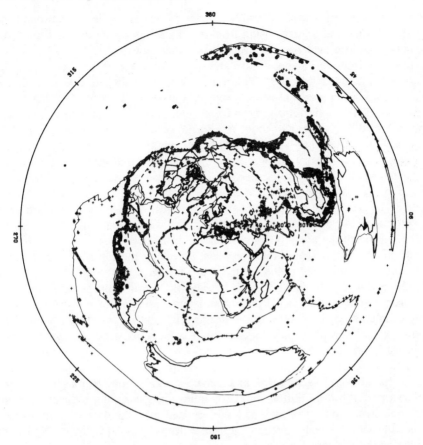

Fig. 2. Azimuthal, equidistant projection centred on Lucerne, Switzerland. Earthquake epicentres (magnitude greater than 4·5) for 1968–1973 are shown by + for shallow earthquakes (depth < 100 km) and by ● for deep earthquakes (depth > 100 km). If there are too many earthquakes in a particular area, they have been omitted for the sake of clarity (courtesy E. R. Kanasewich)

Later large arrays came into operation and results from them posed additional problems, in that the angle of approach predicted by classical methods was often up to 5° away from the measured angle. Davies and Sheppard (1972) found that the anomalies could not be explained satisfactorily without resort to marked lateral heterogeneity deep in the mantle. Julian and Sengupta (1973) then re-examined travel-time data. When every care was taken to avoid lateral heterogeneity in the upper mantle by using only very deep earthquakes as sources, and when the data were (as far as possible) not bunched together in a computer, from which only a mean could be extracted, significant patterns emerged. By retaining a knowledge of the approximate path sampled by each data point, it was possible to see broad general regions in the deep mantle (mainly within a few hundred km of the MCB) from which P waves emerged up to one sec early or late relative to standard travel-time tables. Much of what in earlier studies had been regarded as general scatter arising from reading errors and crustal structure was probably genuine evidence for deep lateral heterogeneity.

Evidence from seismic body wave and surface wave data has indicated that the Earth's upper mantle (depth < 700 km) is strongly laterally heterogeneous: lateral variations of compressional P wave velocities as large as ten per cent have been reported for the upper 200 km, and shear wave velocities probably vary even more. At depths greater than about 700 km, however, it has been more difficult to establish the existence of lateral variations, although such variations have been invoked by various workers (Toksöz *et al.*, 1967; Hales and Roberts, 1970) to explain the scatter of some of the seismological data. Many anomalies have been found which are too large to be the result of effects of upper mantle heterogeneities in the source region; on the other hand, anomalies often vary rapidly with the direction of approach of the waves, implying that structure directly beneath the array is not responsible. Further evidence has come from a study of the diffraction of compressional waves by the Earth's core, in which Alexander and Phinney (1966) found that the region of the MCB beneath the Pacific Basin is different from that beneath the North Atlantic and Africa.

Julian and Sengupta (1973) found that lateral variations of P wave velocity in the middle or lower mantle are necessary to explain travel-time anomalies. However, because of the uneven distribution of seismological observatories and deep earthquakes, sampling of the mantle by available data is uneven, and it is impossible to determine uniquely the complete three-dimensional velocity structure of the mantle. Fig. 3 (after Julian and Sengupta, 1973) shows regions where paths of observed P waves bottom in the mantle—cross-hatching indicates regions differing significantly from the mean determined from the data. The amount by which the actual velocity in the mantle varies depends upon the size of the regions within which variations occur.

Some of the anomalies at the MCB are small-scale: evidence from earthquakes in Tonga to the Solomon Islands suggests scale lengths as small as 200 km or less. On the other hand, evidence from Eurasian earthquakes reveals broader anomalies of scale: perhaps 1000 km. It is interesting to speculate whether these variations are related in some way to convection plumes which Wilson (1965) and Morgan (1971) suggested may extend into the deep mantle—the question of the region beneath Hawaii will be discussed later. Davies and Sheppard (1972) found that rays from the three azimuths of rapid contrast ($20°$, $155°$ and $265°$) bottoming at distances from LASA of about $48°$, come within 500 km laterally of passing through the projection of Iceland, the Galapagos Islands and Hawaii on the MCB. Morgan (1971) had proposed that these surface features may have a source deep in the mantle—the downward extension of 'hot spots'. Except for the Hawaiian Islands, no geological or tectonic features show an obvious correlation with the inferred deep mantle velocity anomalies, although this is not true at shallower depths. The data do not support any correlation between velocity variations below 2000 km and global gravity anomalies or geoid heights. Again at shallower depths such a correlation does exist.

Fig. 3. Regions where paths of observed P waves bottom in the mantle. Cross-hatching indicates regions differing significantly from the mean determined from all data. ▨, Late; ▨, Early (after Julian and Sengupta, 1973)

It is worth while discussing the Hawaiian anomaly in a little more detail. Kanasewich *et al.* (1972), from a study of P wave arrivals at the seismic array VASA in central Alberta, found that waves arriving from earthquakes near the Tonga Islands have anomalous phase velocities. At epicentral distances of 84° phase velocities were normal, but at 95° they were as much as 15 per cent higher than expected. By contrast, no such anomalies were observed from other sources beneath the Pacific, South America or Asia even though they are at comparable distances.

Kanasewich *et al.* ruled out inhomogeneities in the source region and crustal structure beneath the receivers as the cause of the anomalies and believed that they lay at the turning point of the ray in the lower mantle under Hawaii. It is unlikely that the anomalies are due to bumps at the MCB. Buchbinder (1972) found that multiple reflections within the core (P4KP and P7KP) have travel-times that agree very closely with predictions from existing standard Earth models. This is a highly sensitive measure of variations in the core radius—it is unlikely that there are any bumps on the MCB more than a few kilometres in height, whereas the anomalous phase velocities observed would require a depression of 125 km in the surface of the core. Kanasewich *et al.* thus believe that the anomalies are caused by a laterally and radially inhomogeneous region beneath Hawaii.

Lateral limitations may be placed on the size of the inhomogeneity from the fact that shocks from Fiji and from the southern end of the Tonga region give waves with normal phase velocities. In the vertical direction, the data offer no evidence for an extension of the inhomogeneity above a radius of 4000 km. On the other hand, such a possibility is not excluded. Wilson (1963) hypothesized that lava arose from a mantle source within the relatively stagnant centre of a convection cell. The volcanoes in the Hawaiian archipelago would become extinct from northwest to southeast as the Pacific plate moved over the stationary source and such a trend has been confirmed by K–Ar dating (McDougall, 1964). The asthenosphere is an unlikely source for the lavas in the light of plate tectonics and, from seismicity studies on the depth of earthquakes and seismic observations of mantle structure, there is no indication of a source in the mesosphere.

The evidence of an inhomogeneity at the base of the mantle is thus not inconsistent with Wilson's hypothesis for the origin of the Hawaiian Islands and also with Mòrgan's (1972) proposal for the existence of hot spots in the lower mantle. Presentation of the results from several arrays is difficult and Kanasewich and Gutowski (1975) have developed a new method of displaying phase velocities and azimuthal deviations from a great circle path. This involves plotting the data at the turning points of the rays (Fig. 4). The turning points are contoured in terms of anomalous phase velocity (observed minus Jeffreys–Bullen). The Hawaiian Islands have been projected to the level of the turning points which vary from 50 to 300 km above the MCB. The VASA points are on average 100 km above those observed by the LASA array. Since the results from both arrays contour in a simple manner, contain azimuthal deviations that are anomalous, and both sets of rays only intersect near the turning point, the simplest assumption is that the inhomogeneous area occurs near the turning points. It must be pointed out,

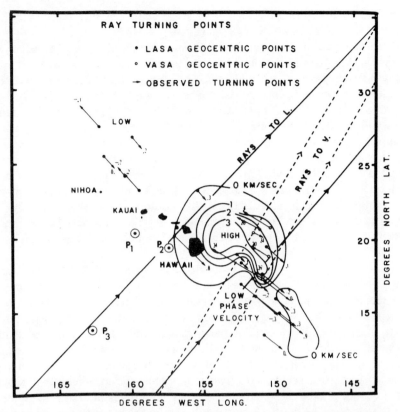

Fig. 4. Contour diagram of anomalous (observed minus Jeffreys–Bullen) phase velocities for earthquakes in the southwest Pacific as recorded by LASA and VASA arrays. The tail of the arrow represents the geocentric location of the turning point on a great circle path between the source and the receiver. The arrowhead represents the actual turning point based on the observed azimuthal deviation. A map of the Hawaiian Islands is projected to the level of the turning points which are just above the core–mantle interface. The diagonal solid lines labelled 'rays to L.' represent two possible great circle paths to LASA. Points P_1, P_2 and P_3 correspond to the closest data points of Wright (1973) and have normal velocities (after Kanasewich *et al.*, 1975)

however, that this interpretation by Kanasewich *et al.* has been criticized by Wright (1975). Wright believes that such anomalies are caused mainly by lateral variations in the crust and upper mantle beneath the arrays and cannot be attributed completely to any simple localized cause. The ray paths from Hawaii pass through the upper mantle beneath the Canadian cordillera where strong lateral variations close to where the Pacific plate abuts the North American plate are to be expected. He also argues that if the assumptions made by Kanasewich *et al.* are applied to measurements at the Warramunga array in Australia or the Yellowknife array in Canada, an unrealistic and unacceptable picture of the Earth's structure is obtained. Kanasewich *et al.* (1975) have attempted to answer some of Wright's criticisms, but it must be confessed that the correct interpretation of the observations is still in doubt.

A new method of studying lateral heterogeneities in the deep mantle has been suggested by Jordan and Lynn (1974). On the seismic trace of a large earthquake many different signals can be identified and timed to a fraction of a second. In addition to the P and S waves, phases called PcP and ScS are clearly visible—they are reflections of P and S waves from the MCB. If the time interval between the phases P and PcP is measured, it should, in a laterally homogeneous Earth, depend only on source–receiver distance. Furthermore, any deviations in elastic properties in the source or receiver regions will be sampled almost equally by the two phases, which follow paths in the upper mantle that are relatively close to each other. Thus to a first approximation upper mantle anomalies are eliminated when the difference of times is taken. At distances greater than about 30°, the differential travel-times PcP–P and ScS–S are not much affected by even gross velocity variations in the crûst and uppermost mantle, and they are also relatively insensitive to event mislocations as well.

Jordan and Lynn carried out a study of ScS–S and PcP–P differential travel-times and absolute S wave travel-times from two deep-focus South American earthquakes and concluded that the scatter observed in ScS–S times by Hales and Roberts (1970) and Jordan (1972) resulted from lateral heterogeneities in the lower mantle along the path of the S phase. Both ScS–S and PcP–P times delineate a region of anomalously high velocity in the lower mantle beneath the Caribbean. They estimated the total velocity variation associated with this anomaly to be one per cent or greater. A velocity heterogeneity that is purely radial cannot explain the scatter in the observed differential travel-times—much of the scatter in the residuals must be due to lateral variations in the velocities. The observed velocity contrasts may be due to a thermal anomaly, in which case lateral temperature differences of at least 200–300°C are required. Jordan and Lynn suggested that the high-velocity anomaly beneath the Caribbean marks a site of descending material in the convecting mantle.

Reversals of the Earth's Magnetic Field and the MCB

Fig. 5 shows the average frequency of reversals of the Earth's magnetic field seen in a sliding window 10 Ma long. Detailed studies of selected segments of the time scale indicate that the true frequency may be as much as 50 per cent greater. It can be seen that the mean frequency of reversals has remained practically constant for the past 45 Ma. The mean frequency of reversals shows no statistically significant changes from about 75–45Ma ago and then increases by a factor of more than two. This comparatively sudden increase in reversal frequency was first noticed by Heirtzler et al. (1968). An age of 45 Ma thus appears to mark a boundary between two intervals during which the statistical properties of the geodynamo were very different. An even more striking 'discontinuity' occurred about 85 Ma ago. For the previous 22 Ma, there appear to have been no reversals (the Cretaceous normal polarity interval). Thus during the past 100 Ma, the fre-

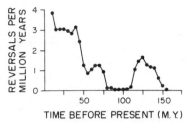

Fig. 5. Variations in the average frequency of reversals as seen through a sliding window 10 Ma long. From 76 Ma to the present the rates are from Heirtzler *et al.* (1968) and prior to 76 Ma from Larson and Pitman (1972) (after Cox, 1975)

quency of reversals increased from zero to the present frequency of about five per Ma in two steps, one at 85 Ma and one at 45 Ma. There have also been other instances of increases and decreases in the reversal frequency in the past (see for example McElhinny, 1971) changes in reversal frequency typically occurring at time intervals of about 50 Ma.

The characteristic time associated with an individual reversal (0·2 Ma) is orders of magnitude different from that associated with changes in reversal frequency. It seems very probable that the two phenomena have different causes. Individual reversals are more likely to be the result of fluid motions and electric currents in the core. Changes in reversal frequency on the other hand are more likely to be due to changes in the rate at which energy is available to generate turbulence in the core or to a change in the boundary conditions at the MCB. The characteristic times of processes that could affect the energy supply are not easy to estimate, but are likely to be comparable with the age of the Earth. A time period of the order of 50 Ma on the other hand is of the same order as that associated with geological events—the formation of ocean basins and mountain ranges. It is also a reasonable time to associate with convection in the mantle. The basic question then is whether convection takes place throughout the mantle and, if so, whether it could affect the characteristics of the geodynamo. It is possible that deep mantle convection could cause hot and cold spots at the MCB and/or bumps on the MCB: for example a cold descending column could produce both a cold spot and a bump due to the greater density of the material in the column.

There is some evidence for a correlation between major tectonic changes and changes in reversal frequency. Fig. 6 shows the magnetic reversal chronology for the last 80 Ma laid out alongside the Hawaii–Emperor chain. It can be seen that there is a major change in direction of the Pacific/mantle motion which occurred around 42–44 Ma ago, producing the Hawaii–Emperor Bend. Moreover a pronounced change in azimuth of the Mendocino fracture zone occurred about 42 Ma ago reflecting a reorientation in spreading between the Pacific and Farallon plates. These dates coincide with the major change in reversal frequency around 45 Ma within the limits of probable error. Again 42–45 Ma marks the

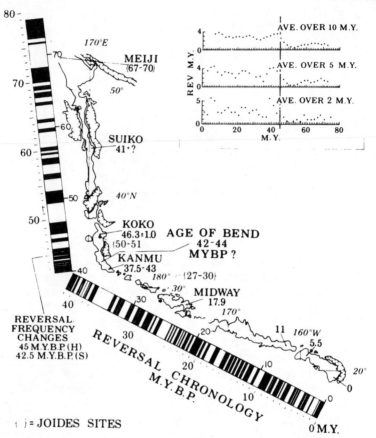

Fig. 6. Age dates along the Hawaii–Emperor chain and magnetic reversal chronology. The age scale along the lower edge is from Heirtzler *et al.* (1968) and that along the upper edge from Sclater *et al.* (1974). The smoothed reversal frequency (Heirtzler *et al.*, 1968) is shown in the upper right-hand corner with the solid vertical line marking the time of the major reversal frequency change about 45 Ma B.P. which is also about the age of the Hawaii–Emperor Bend. The revision of Heirtzler *et al.*'s time scale by Sclater *et al.* moves the frequency change from 45 to 42·5 Ma B.P. (after Vogt, 1975)

largest change in direction of plate motion between Greenland and Europe (Vogt and Avery, 1974). Vogt (1975) has also shown that major tectonic changes occurred about 77 and 115 Ma ago, coinciding within the dating uncertainty with reversal frequency changes.

Hide (1969) had suggested earlier that irregular features at the MCB might provide topographic coupling between the core and mantle and account for the decade fluctuations in the length of the day. Bumps on the MCB may also be the cause of horizontal density variations responsible for regional gravity anomalies. It can readily be shown (for example Hide and Horai, 1968) that, because of the density contrast at the MCB, bumps with horizontal dimensions up to thousands of kilometres and a kilometre or so in height would make a significant (although not dominant) contribution to the observed distortion of the gravitational field at the Earth's surface. Hide (1967, 1969, 1970) also suggested that bumps on the MCB might affect the flow pattern in the core and thus influence the detailed configuration of the geomagnetic field and its time variations. While the liquid core of the Earth is the only likely location of electric currents responsible for the main geomagnetic field, it is the most unlikely place to find density variations of sufficient magnitude to cause the observed distortions of the gravitational field—these must arise largely in the mantle. Thus any correlation between gravity and magnetic anomalies should reflect processes at the MCB. Hide and Malin (1970) argued that if both gravity and magnetic anomalies are the result of the same topographic features, it should be possible to find a statistically significant correlation between them. In fact they found, for spherical harmonic coefficients up to degree 4, a correlation coefficient of $0 \cdot 84$ between large-scale features of the Earth's non-dipole magnetic field (for epoch 1965) and the gravitational field, provided the magnetic field is displaced 160° eastward in longitude λ. Hide and Malin also showed that λ has increased linearly with time since 1835, the date of the earliest reliable spherical harmonic analysis (by Gauss) of the geomagnetic field. This dependence of λ on time is associated with the westward drift of the geomagnetic field. It must be stressed that even if the correlation does exist (as seems most likely) it does not by itself prove the existence of bumps on the MCB. It is possible that quite small temperature variations over the MCB could, through their effects on core motions, produce measurable distortions of the geomagnetic field. If these temperature variations in turn reflect the density structure of the lower mantle, then there would be a correlation between gravity and geomagnetic anomalies.

Robinson (1974) has developed a boundary-layer model of thermal convection throughout the Earth's mantle and been able to estimate the distortion of the MCB due to such convective motion. By equating the additional gravitational force of the heavier descending plume with the hydrostatic force due to the distortion of the MCB, he estimated the displacement to be about $1 \cdot 5$ km, which is of the order of that required by Hide (1969, 1970) to account for core–mantle coupling.

The Thermal Regime of the Core

For a proper understanding of the thermal regime of the Earth's core, it is necessary to determine the adiabatic and melting (or liquidus) temperatures. If the transition from the inner to the outer core is a transition from the solid to the liquid form of a single material, then the boundary must be at the melting point and a constraint is put on the thermal regime of the Earth's interior. Jacobs (1953) used this fact to explain how the mantle and inner core could be solid, while at the same time the outer core is liquid.

In his discussion, no specific values of the temperatures are postulated and the behaviour of the adiabatic and melting point curves need not be known precisely;

it is necessary only that the adiabatic gradient in the outer core be less than the melting point gradient. However, in 1971, Higgins and Kennedy found that the adiabatic gradient in the outer core was steeper than the melting point gradient, in which case the argument put forward by Jacobs would no longer be valid.

If actual temperatures were distributed along the adiabat of Higgins and Kennedy throughout the outer core, it too would be solid: there would be no liquid outer core (see Fig. 7). Higgins and Kennedy thus concluded that the actual temperature gradient in the outer core is much less than the adiabatic gradient. If this is the case the outer core would be thermally stably stratified, thereby inhibiting radial convection which is necessary to drive the geodynamo.

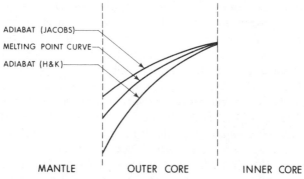

Fig. 7. Melting point curve of iron and adiabatic temperature curves of Jacobs (1971) and Higgins and Kennedy (1971) versus depth in the Earth's outer core

There have been a number of attempts to resolve what has been called the 'core paradox'. In the first place it must be stressed that iron exists in several crystalline forms and that the melting curve defines a condition of equilibrium between the liquid phase and one of the solid phases. Four crystalline phases of iron are known (α, γ, δ and ϵ). Only at low pressures is the α phase in equilibrium with the melt. Our knowledge of the thermodynamic properties of the γ and ϵ phases (of most geophysical interest) is very rudimentary. Birch (1972) estimated that the γ melting temperatures are about 700°C higher than those of Higgins and Kennedy, with ϵ melting temperatures still higher; he does not believe that present evidence is sufficient to predict melting temperatures of iron at core pressures to within 500°C. Again it must not be forgotten that the core is not pure iron, but contains about 15 per cent of some light alloying element (probably Si or S). This would modify any phase relationships, and also considerably lower melting temperatures in the outer core—Hall and Murthy (1972) suggested that the eutectic temperature may be some 1600°C lower than that for pure iron. There is also evidence that the depth of the eutectic trough deepens with increasing pressure (Kim *et al.*, 1972). Finally, as Verhoogen (1973) has pointed out, the melting point curve of Higgins and Kennedy is based on a 100-fold linear extrapolation of 30 kbar experimental results. The adiabatic temperature gradient can only be estimated theoretically and depends quite critically on the values of certain parameters in the core which are but imperfectly known.

The immediate reaction to Kennedy and Higgins' paper was to accept their results at face value and try to invent ways and means to get around the core paradox. Thus Bullard and Gubbins (1971) showed that internal waves represent the only motions with a significant radial component in the case of stratification. They did not, however, offer any suggestion for an energy source to produce such oscillatory motion. Busse (1972), Malkus (1973) and Elsasser (1972) independently suggested that the outer core might consist of a slurry of fine iron particles suspended in an iron-rich liquid. There is however a critical limit to the solid grain size in order that they do not precipitate out faster than the core can stir them up. Malkus (1973) estimated that the critical size is one micron for a convection-driven dynamo and ten microns for a precession-driven dynamo. Such small grain sizes do not seem to be in accord with metallurgical experience. Bukowinski and Knopoff (1973) and Stacey (1975) suggested that the inner core may be an electronic phase transition in iron to the $3d^8$ state, that is, a collapse of all of the conduction or valence electrons into vacant 3d states. The inner core would then be a dense liquid or glassy phase of iron and this would remove the constraint on the temperature of the core, based on the assumption that the inner core boundary represents the intersection of the temperature profile with the solidus of an iron-rich mix. Finally, Schloessin (1974) has suggested that constitutional supercooling (that is, crystallization from impure solutions or melts) might be important in the evolution of the core. A slow overgrowth of the mantle and inner core at the expense of the outer core might be a continuing process and thereby give rise to bumps on the MCB. By this means density inhomogeneities could be generated independent of thermal inhomogeneities—even in a predominantly stratified core.

Jacobs (1971) and Birch (1972) have re-estimated the melting-point and adiabatic gradients in the core and find that the adiabatic gradient is less (slightly) than the melting-point gradient. However, in view of all the uncertainties in the estimates of these gradients, it is impossible to say with any conviction which is the steeper in the core. I personally believe that actual temperatures in the core are probably very close to the melting temperature. If the melting-point and adiabatic gradients are virtually the same throughout the entire core, then perhaps parts of the mantle may from time to time become soluble in the outer core or material from the outer core diffuse into the lower mantle: thereby giving rise to bumps on the MCB. The position and shape of the boundary may thus change over the course of time and be instrumental in initiating and dictating motions in the outer core. If the topography of either boundary of the outer core has a direct influence on core motions, then changes in the frequency of reversals of the Earth's magnetic field may well be random as the shape of these boundaries (randomly) changes.

If the temperatures in the outer core are very close to the melting temperature, convection may only just be possible. This would enhance the inherent instability of the dynamo process and affect the relative frequency of reversals in the recent past. The whole regime of reversals, their frequency, duration and relative times of different polarity, appears to be determined by the parameters of the Earth's core and hence, as the core evolves, both the regime of reversals and their frequency will change. It is known that the secular variation over the central Pacific has been systematically weaker than elsewhere for several million years (see for example Doell and Cox, 1972). It is interesting to speculate whether this is caused by the interaction of core motions with bumps on the MCB.

There seems to be little doubt that lateral inhomogeneities exist in the deep mantle and I suspect that there is topography on the MCB. Present data however are insufficient to make definite statements about the deep interior of the Earth beneath Europe.

References

Alexander, S. S. and Phinney, R. A. (1966). 'A study of the core–mantle boundary using P waves diffracted by the Earth's core', *J. Geophys. Res.*, **71**, 5943

Birch, F. (1972). 'The melting relations of iron and temperatures in the Earth's core', *Geophys. J.*, **29**, 373

Buchbinder, G. G. R. (1972). 'Travel-times and velocities in the outer core from PmKP', *Earth Planet. Sci. Letters*, **14**, 161

Bukowinski, M. and Knopoff, L. (1973). *Electronic transition in iron and the properties of the core*. Publ. No. 1362, Inst. Geophys. Planet. Phys., Univ. Calif., Los Angeles

Bullard, E. C. and Gubbins, D. (1971). 'Geomagnetic dynamos in a stable core', *Nature*, **232**, 548

Busse, F. H. (1972). Comment on the paper 'The adiabatic gradient and the melting point gradient in the core of the Earth' by Higgins, G. and Kennedy, G. C., *J. Geophys. Res.*, **77**, 1589

Cox, A. (1975). *The frequency of geomagnetic reversals and the symmetry of the non-dipole field*, IUGG Quadrennial Report for the USA

Davies, D. and Sheppard, R. M. (1972). 'Lateral heterogeneity in the Earth's mantle', *Nature*, **239**, 318

Doell, R. R. and Cox, A. (1972). 'The Pacific geomagnetic secular variation anomaly and the question of lateral uniformity in the lower mantle', in Robertson, E. C. (ed.), *The Nature of the Solid Earth*, McGraw-Hill

Elsasser, W. M. (1972). 'Thermal stratification and core convection', Int. Conf. Core–Mantle Interface, E S, *Trans. Amer. Geophys. Union*, **53**, 605

Hales, A. L. and Roberts, J. L. (1970). 'Shear velocities in the lower mantle and the radius of the core', *Bull. Seism. Soc. Amer.*, **60**, 1427

Hall, H. T. and Murthy, V. R. (1972). 'Comments on the chemical structure of an Fe–Ni–S core of the Earth', Int. Conf. Core–Mantle Interface, E S, *Trans. Amer. Geophys. Union*, **53**, 602

Heirtzler, J. R., Dickson, G. O., Herron, E. M., Pitman, W. C. III and Le Pichon, X. (1968). 'Marine magnetic anomalies, geomagnetic field reversals and motions of the ocean floor and continents', *J. Geophys. Res.*, **73**, 2119

Hide, R. (1967). 'Motions of the Earth's core and mantle and variations of the main geomagnetic field', *Science*, **157**, 55

Hide, R. (1969). 'Interaction between the Earth's liquid core and solid mantle', *Nature*, **222**, 1055

Hide, R. (1970). 'On the Earth's core–mantle interface', *Q. J. Met. Soc.*, **96**, 579

Hide, R. and Horai, K.-I. (1968). 'On the topography of the core–mantle interface', *Phys. Earth Planet. Int.*, **1**, 305

Hide, R. and Malin, S. R. C. (1970). 'Novel correlations between global features of the Earth's gravitational and magnetic fields', *Nature*, **225**, 605

Higgins, G. H. and Kennedy, G. C. (1971). 'The adiabatic gradient and the melting-point gradient in the core of the Earth', *J. Geophys. Res.*, **76**, 1870

Jacobs, J. A. (1953). 'The Earth's inner core', *Nature*, **172**, 297

Jacobs, J. A. (1971). 'The thermal regime in the Earth's core', *Comments Earth Sci. Geophys.*, **2**, 61

Jordan, T. H. (1972). *Estimation of the radial variation of seismic velocities and density in the Earth*, Ph.D. Thesis, Cal. Inst. Tech., Pasadena

Jordan, T. H. and Lynn, W. S. (1974). 'A velocity anomaly in the lower mantle', *J. Geophys. Res.*, **79**, 2679

Julian, B. R. and Sengupta, M. K. (1973). 'Seismic travel-time evidence for lateral inhomogeneity in the deep mantle', *Nature*, **242**, 443

Kanasewich, E. R., Ellis, R. M., Chapman, C. H., and Gutowski, P. R. (1972). 'Teleseismic array evidence for inhomogeneities in the lower mantle and the origin of the Hawaiian Islands', *Nature Phys. Sci.*, **239**, 99

Kanasewich, E. R. and Gutowski, P. R. (1975). 'Detailed seismic analysis of a lateral mantle inhomogeneity', *Earth Planet. Sci. Letters*, **25**, 379

Kanasewich, E. R., Ellis, R. M., Chapman, C. H. and Gutowski, P. R. (1975). Reply to comments on 'Seismic array evidence of a core boundary source for the Hawaiian linear volcanic chain' by Kanasewich, E. R. *et al.*, *J. Geophys. Res.*, **80**, 1920

Kim, Ki-Tae, Vaidya, S. N. and Kennedy, G. C. (1972). 'Effect of pressure on the temperature of the eutectic minimums in two binary systems: NaF–NaCl and CsCl–NaCl', *J. Geophys. Res.*, **77**, 6984

Larson, R. L. and Pitman, W. C. III (1972). 'World-wide correlation of Mesozoic magnetic anomalies and its implications', *Bull. Geol. Soc. Amer.*, **83**, 3645

Malkus, W. V. R. (1973). 'Convection at the melting point, a thermal history of the Earth's core', *Geophys. Fluid Dyn.*, **4**, 267

McDougall, I. (1964). 'Potassium-argon ages from lavas from the Hawaiian Islands', *Bull. Geol. Soc. Amer.*, **75**, 107

McElhinny, M. W. (1971). 'Geomagnetic reversals during the Phanerozoic', *Science*, **172**, 157

Morgan, W. J. (1971). 'Convection plumes in the lower mantle', *Nature*, **230**, 42

Morgan, W. J. (1972). 'Deep mantle convection plumes and plate motions', *Amer. Assoc. Petrol. Geol. Bull.*, **56**, 203

Needham, R. E. and Davies, D. (1973). 'Lateral heterogeneity in the deep mantle from seismic body wave amplitudes', *Nature*, **244**, 152

Phinney, R. A. and Alexander, S. S. (1969). 'The effect of a velocity gradient at the base of the mantle on diffracted P waves in the shadow', *J. Geophys. Res.*, **74**, 4967

Robinson, J. L. (1974). 'A note on convection in the Earth's mantle', *Earth Planet. Sci. Letters*, **21**, 190

Schloessin, H. H. (1974). 'Corrugations on the core boundary interfaces due to constitutional supercooling and effects on motion in a predomiantly stratified liquid core', *Phys. Earth Planet. Int.*, **9**, 147

Sclater, J. G., Jarrad, R., McGowran, B. and Gartner, S. Jr. (1974). In van der Borch, C. C., Sclater, J. G. and others, *Initial Reports of the Deep Sea Drilling Project*, **22**, U.S. Government Printing Office, Washington, DC

Stacey, F. D. (1975). 'Thermal regime of the Earth's interior', *Nature*, **255**, 44

Toksöz, M. N., Chinnery, M. A. and Anderson, D. L. (1967). 'Inhomogeneities in the Earth's mantle', *Geophys. J.*, **13**, 31

Verhoogen, J. (1973). 'Thermal regime of the Earth's core', *Phys. Earth Planet. Int.*, **7**, 47

Vogt, P. R. (1975). 'Changes in geomagnetic reversal frequency at times of tectonic change: evidence for coupling between core and upper mantle processes', *Earth Planet. Sci. Letters*, **25**, 313

Vogt, P. R. and Avery, O. E. (1974). 'Detailed magnetic survey in the northeast Atlantic and Labrador Sea', *J. Geophys. Res.,* **79**, 363

Wilson, J. T. (1963). 'A possible origin of the Hawaiian Islands', *Can. J. Phys.,* **41**, 863

Wilson, J. T. (1965). 'Convection currents and continental drift', *Phil. Trans. Roy. Soc.,* **258A**, 145

Wright, C. (1973). 'Observations of multiple core reflections of the PnKP and SnKP type and regional variations at the base of the mantle', *Earth Planet. Sci. Letters,* **19**, 453

Wright, C. (1975). Comments on 'Seismic array evidence of a core boundary source for the Hawaiian linear volcanic chain', by Kanasewich, E. R. *et al., J. Geophys. Res.,* **80**, 1915

Eo-Europa

That part of the continent unaffected by
major tectonism since Precambrian times

EO-EUROPA: THE EVOLUTION OF A CRATON

JANET WATSON

Department of Geology, Imperial College, London, U.K.

Abstract

Watson, J. (1976). '*Eo-Europa:* the Evolution of a Craton', in Ager, D. V. and Brooks, M. (eds.), *Europe from Crust to Core*, Wiley, London.
The geological history of *Eo-Europa* falls into two stages of roughly equal length, the earlier (more than 1800 Ma) being one of widespread crustal mobility and the later one of relative stability. The early stage recorded in the crystalline basement led first to the formation of Archaean complexes, consisting largely of gneisses and granites, and subsequently to the development of early Proterozoic belts enclosing small massifs of little-modified Archaean rocks. The well-known distinctions between Proterozoic regions such as the Karelide subprovince which incorporates an Archaean basement and the Svecofennide subprovince which does not, point to a contrast between possible ensialic and ensimatic mobile belts. The mineralized junction-zone of the Karelide and Svecofennide regions coincides with a northwesterly lineament belonging to a set widely developed in later Proterozoic times.
The main part of *Eo-Europa* was established as a continental craton in Mid-Proterozoic times. Its subsequent development reflects both the operation of the internal cratogenic regime and the effects of external events connected with the alteration of phases of disruption and extension of continental masses with phases of collision and orogeny. During its initial phases the eocraton probably formed part of a very large stable unit. A unique event of this stage was the emplacement of a rapakivi granite/anorthosite suite in a zone which probably extended to Greenland and North America.
Many of the distinctive features of *Eo-Europa* appear to have been established at a remarkably early stage. Lateral variations in the thickness of the crust and the density of lower crustal and/or upper mantle rocks may have been preserved since Mid-Proterozoic times. The broad distinction between shield and platform areas and between regions affected by strong block-faulting and more coherent regions may date from a similar time. Old lineaments have been repeatedly reactivated. Alkaline and basic igneous bodies have been emplaced over a wide time-span although the general level of cratonic igneous activity remained consistently low.

The Geological Context

That portion of the European craton which Stille termed *Eo-Europa* has remained free from orogenic disturbances for more than 600 Ma: much of it indeed has been more or less stable for 1700 Ma. It is flanked on the east, south and northwest by regions which were stabilized only after the end of Palaeozoic orogenic cycles and were welded to it between 400 and 200 Ma ago. The aim of this paper is not to summarize the structure and geological history of *Eo-Europa*, which are well known through the work of generations of European geologists and geophysicists, but to examine some

59

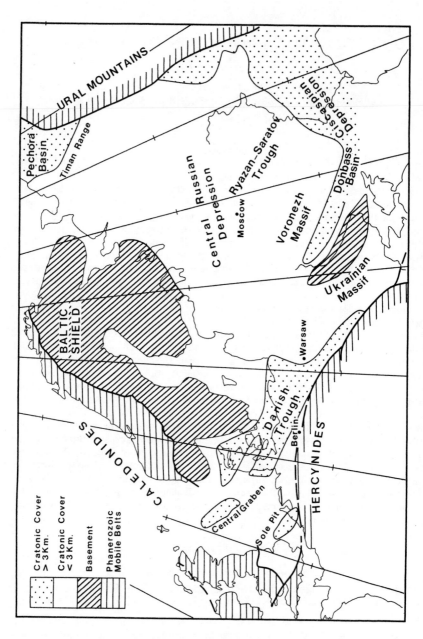

Fig. 1. Locality map of *Eo-Europa* showing marginal relationships and distribution of cratonic cover

features which bear on the evolution of this long-lived craton and on its relationship with the younger tectonic provinces at its margins.

The main mass of *Eo-Europa* has an area of some 4.5×10^6 km^2 of which less than a quarter is occupied by outcrops of the Precambrian basement and the remainder by cratonic cover-formations ranging in age from Mid-Proterozoic to Quaternary. A much smaller cratonic unit, separated from the main area by the Caledonian tectonic province, occupies the northwestern portion of the British Isles. This Hebridean craton, though geographically a part of Europe, is geologically related to the Greenlandic and North American cratons from which it has been detached by the opening of the Atlantic Ocean.

The main portion of *Eo-Europa* (Fig. 1) is remote from active plate margins. It comes directly in contact with the Alpine–Himalayan orogenic belt only along part of the Carpathian arc and is elsewhere separated from this belt by stabilized portions of the Hercynian tectonic province. It does not extend to the Atlantic margin and may also be separated from the Arctic oceanic basin by a Palaeozoic tectonic province. The small Hebridean craton differs from it in that it is directly truncated by the continental slope of the North Atlantic which originated as a fracture-coastline in Mesozoic times. Apart from the vertical movements associated with recent glaciation, geological activity in *Eo-Europa* has been slight for at least the past 100 Ma. Deep basins containing Tertiary and Mesozoic sediments are almost confined to the border regions, as are topographical highlands. Igneous activity has been limited since the end of the Palaeozoic era. Heat-flow is low (figures quoted by Belyaevsky *et al.* (1973) are seldom more than 1.0 μm cal/cm^2 s^{-1}) and suggests a geothermal gradient giving temperatures of only 250–400°C at the base of the crust. Current seismic activity is very slight and is almost confined to old-established lineaments. In all these respects *Eo-Europa* appears to be behaving as a mature or senescent crustal unit.

A very large part of the basement of *Eo-Europa*, together with that of the Hebridean craton, was stabilized at about 1800 Ma and suffered no subsequent disturbances of orogenic character. I shall therefore adopt a date of about 1800 Ma—the time at which the Svecofennide and Karelide cycles came to an end—as that at which the forerunner of the *Eo-European* craton came into existence as a geological entity. The dominant crustal regimes which operated on a regional scale before and after this date were of course very different. The effects of the early regimes are recorded in the older portions of the crystalline basement, those of the later regimes in both basement and cratonic cover.

Early History of the Crystalline Basement

In the basement of the embryonic European craton a number of ovoid Archaean provinces more than 2500 Ma in age are enclosed within a more extensive terrain which is characterized by early Proterozoic (older than

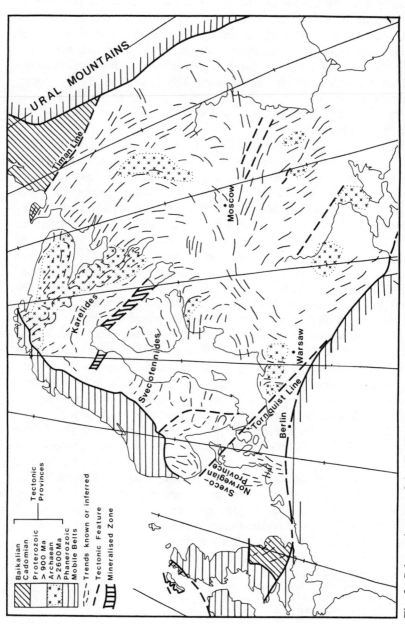

Fig. 2. Schematic map of the *Eo-European* craton. Tentative trend-lines in the basement beneath the cratonic cover are derived from published sources, especially the Tectonic Map of Europe (1964), the Basement Map of the USSR (1974) and the Aeromagnetic Map of Britain (1965). The Archaean areas distinguished represent massifs in which Proterozoic tectonic effects are slight: regenerated Archaean rocks in Proterozoic provinces are not shown

1800 Ma) tectonic patterns (Fig. 2). The Archaean massifs are small when compared with such Archaean units as the Superior province of Canada, that forming the northeastern part of the Baltic Shield, for example, having a long diameter of about 800 km. In this massif, in those of the Ukraine and, so far as is known, in subcrops of Archaean beneath the cratonic cover, the dominant rocks are granites, gneisses and low-pressure granulites associated with metasediments and metavolcanics of medium or high metamorphic grade. The scarcity on the one hand of low-grade greenstone-belts with their characteristic gold mineralization and on the other of the layered anorthositic complexes typical of many deep-level Archaean gneiss and granulite terrains produces a broad uniformity of character which suggests that all the Archaean massifs of *Eo-Europa* may have been eroded to roughly the same depth. Archaean remnants incorporated in the early Proterozoic province of the Hebridean craton, on the other hand, include medium- to high-pressure granulites with layered anorthositic bodies like those of the pre-Ketilidian massif of Greenland and may reveal a significantly different crustal assemblage.

The Belamoride belt which forms a distinctive NW–SE tract in the Archaean portion of the Baltic Shield has commonly been regarded as a late Archaean or early Proterozoic structure. Zircon and other dates of 2800–2600 Ma have, however, been recorded in addition to a large number of younger mineral ages and suggest that the belt may be as old as the adjacent rocks (Kratz *et al.*, 1968). It is analogous in some respects to the Limpopo belt of southern Africa which represents a highly deformed and strongly elevated Archaean tract between two granite–greenstone belt provinces. The Lapland granulites may mark the site of a very ancient lineament (*fracture de fond* of Kratz *et al.*, 1974).

The early Proterozoic tectonic provinces which form much of the *Eo-European* basement enclose the Archaean massifs in a characteristic mesh-structure. On a gross scale the Early Proterozoic belts appear broad relative to their length and their boundaries are sinuous. Their internal tectonic patterns tend to display curved or closed structures rather than simple linear trends.

In a very broad sense the Early Proterozoic terrains of *Eo-Europa* can be assigned to two categories, reflecting the classic separation of Karelide and Svecofennide terrains in the Baltic Shield. Although this separation does not seem to have the time-significance originally assigned to it (for example by Sederholm, 1930) its geological reality does not seem to be in doubt. In the Karelides of the eastern Baltic Shield, in the Ukraine and in much of the intervening platform-area, Early Proterozoic supracrustal sequences are closely associated not only with Proterozoic granites but also with partly regenerated Archaean crystalline complexes which appear to represent a basement of continental type. The supracrustal rocks of these regions include shallow-water sediments among which are orthoquartzites, dolomites and banded iron-formations deposited under relatively stable conditions. These features, with the very wide distribution of reworked Archaean basement-units (for example, see the Basement Map of the USSR

1974), suggest that the Karelide subprovinces may be largely ensialic, though the possibility that they contain concealed sutures is raised by accounts of rocks identified as ophiolites in Karelia (see Khain, this volume). In the Svecofennide subprovince of the Baltic Shield, Early Proterozoic supracrustal assemblages consist almost exclusively of metavolcanics and detrital sediments deposited under unstable conditions and although Proterozoic granites are abundant there is no identifiable basement of continental type. The nature of the supracrustal assemblage and associated ore-deposits bears some resemblance to those of continental margin environments. The copper–nickel(–zinc) deposits of the main sulphide zone in Finland and the copper–zinc–lead deposits of the Skellefte ore-field of Sweden, both of which border the Svecofennides area, have recently been interpreted in terms of mineralization in an island arc environment, for example by Kahma (1973) and Rickard and Zweifel (1975) respectively.

The exposed Svecofennide subprovince has a poorly defined east–west tectonic grain and is roughly equi-dimensional. It is flanked on the northeast by the type Karelide subprovince and on the south by an area incorporating several Archaean basement massifs. Towards the west its marginal relationships, though complicated by the effects of later tectonic events in the Sveconorwegian and Caledonian provinces, suggest that it is bordered here also by regions incorporating Archaean crustal rocks. The extent of the Svecofennide subprovince therefore seems to be limited and tectonic trends in adjacent areas suggest that the subprovince forms a kind of node in the Early Proterozoic structure (Fig. 2). These relationships would be consistent with the idea that it marks the site of a small ocean basin opened within continental crust rather than part of the linear margin of a large continent.

In this context, the nature of the NW–SE line of junction of the Svecofennide and Karelide subprovinces is of interest. This boundary, which I shall refer to as the Karelide lineament, is oblique to the Svecofennide trend and roughly parallel to that of the Karelides. It is emphasized by major gravity and magnetic anomalies and by the zone of mineralization already mentioned. Its role as a boundary between supracrustal assemblages of differing character shows that it dates back at least to the time of deposition of these assemblages (more than 2000 Ma) while its length and the nature of the associated mineralization indicate that it extends in depth to the mantle. Evidence of transcurrent (dextral) movements is seen in Finland (Gaal, 1972) and transcurrent movements of opposite sense appear to be indicated on parallel fractures some distance to the north of the lineament in Sweden (*S.G.U.* Aeromagnetic maps, 1967). Taken together, these features suggest that the Karelide lineament may have functioned in Early Proterozoic times as a transform dislocation on which movements allowing the opening or closing of a small ocean on the site of the Svecofennide subprovince took place. There are possible analogies with northeasterly lineaments of similar age in North America which appear to have originated as transform structures accommodating internal motions within a large moving continent (Sutton and Watson, 1974). The Nelson River lineament of this set is characterized by important nickel mineralization.

Later Additions to the Basement

The tectonic patterns built up during the Archaean and early Proterozoic cycles mentioned above are preserved with little modification over very large areas of the basement in *Eo-Europa*. Effects of later periods of mobility are clearly displayed in a number of marginal regions which were not incorporated into the eocraton until Late Proterozoic times. In addition, some interior regions record Mid-Proterozoic events of anomalous character which are mentioned for convenience in this section.

Near the western edge of *Eo-Europa* the Sveconorwegian province, with a northerly grain, is dominated by the effects of tectonothermal activity over a period terminating at about 1000 Ma. The external relationships of this province, and in particular its connection with the Grenville province of Canada, are obscure. It may underlie parts of Denmark (where gneisses have been found in boreholes on the Fyn–Grinsted 'high'), north Germany and the southern North Sea (where the alignment of salt ridges in the cratonic cover suggests that the basement may have a northerly grain). It does not appear to continue eastward into the main part of *Eo-Europa*. The problem of its origin lies outside the scope of this review.

Tectonic patterns dating from very late Precambrian or even earliest Palaeozoic episodes characterize the Baikalian province of the Timan region and appear (Khain, this volume) to continue in the basement of the Pechora basin. Rocks and structures in the poorly exposed basement of eastern England and the adjacent areas suggest that phases of broadly similar age (the Cadomian phases) were of importance in the westernmost salient of *Eo-Europa* (Khain, this volume). The close connection between Baikalian and Cadomian mobile regions and the Palaeozic system of mobile belts has led to doubts as to whether the Timan region and much of the 'Danish triangle' (Ager, 1975) should be regarded as parts of *Eo-Europa*. I have considered them in this paper because, as will be seen later, their structures appear to be controlled to some extent by older features in the craton.

The interior of the eocraton is traversed by a broad ESE zone characterized by the occurrence of a distinctive igneous suite comprising rapakivi granite, monzonite, anorthosite, gabbro, norite and a variety of volcanic or subvolcanic associates. This suite, taken as a whole, is a well-known one which has a wide geographical distribution but a restricted time-range. The majority of dated bodies in *Eo-Europa* fall in the range 1800–1400 Ma. On a continental reconstruction which restores Europe to a position alongside North America, the zone in which they occur is seen to be continued westward through southern Greenland and eastern Canada to the western states of USA where similar intrusions have a slightly younger age-range. Some genetic connection with the Early Proterozoic (Karelide–Svecofennide–Hudsonian) system of mobile belts is suggested by the fact that although the igneous suite is clearly post-orogenic it is confined to provinces affected by Early Proterozoic tectonism and is virtually absent from Archaean massifs. Bridgwater and Windley (1973), in a survey of the Mid-Proterozoic rapakivi-anorthosite suite as a whole, infer that the anomalous zone in which it was generated was one of mantle upwelling and that it remained a zone of high heat flow for several hundred million years. This inference is supported by the fact that isotopic ages, particularly mineral ages, which have been obtained from the subcrop of the zone beneath the cratonic cover in the USSR tend to scatter over the range 1700–1200 Ma. This spread of ages, combined with evidence suggesting the presence of structural lines discordant to the Karelian tectonic patterns, has led some authors to infer the existence of Mid-Proterozoic (Volynian, Ovruch) mobile belts in the hidden basement between the Ukraine and the Urals (for example Kratz *et al.*, 1968). The alternative, favoured, for example, by Nalivkin (1970, pp. 133–4), that the Post-Karelian structures in this region are essentially cratogenic in character seems to be supported by the results of recent surveys (Khain, this volume; Basement Map of USSR 1974). I suggest below that some Mid-Proterozoic structures are related to lineaments of a set which remained intermittently active over a very long time period.

If the evidence that most of *Eo-Europa* constituted a single cratonic entity throughout Mid-Proterozoic times is combined with that supplied by the apparent continuity of the anomalous zone invaded by the anorthosite–rapakivi suite in Europe, Greenland and North America (Bridgwater and Windley, 1973) and with the provisional results of palaeomagnetic studies (Piper, 1975), one is led to conclude that over the period from about 1800–1100 Ma the eocraton formed part of a very much larger cratonic unit extending to North America. The supercontinental dimensions of such a craton could be matched only at one other period—that immediately preceding the opening of the Atlantic in Mesozoic times. Its life-span would be without parallel in later periods and would suggest the operation of a distinctive mantle regime.

The Architecture of the Craton

The effects of the long period of cratogenic development which followed the ending of the Karelide cycle are shown not only by the stratigraphical and structural record but also by the fundamental architecture of the craton itself. Deep seismic studies completed in connection with the Upper Mantle Project have revealed unexpectedly large variations in crustal thickness (Fig. 3). As Belyaevsky *et al.* (1973) point out, these variations are not correlated with variations in topographical elevation: from the White Sea to the Ukraine, the land-surface seldom rises above 300 m while the Moho ranges in depth from 35–60 km. The results of gravity surveys do not indicate large departures from isostatic equilibrium and, consequently, isostatic balance in the craton must be related in part to lateral variations in the density of crustal and/or mantle rocks. As will be seen later, some of these variations may date from an early stage in the evolution of the craton.

The relief of the Moho bears some relationship to that of the top of the crystalline basement. In the northern and central parts of *Eo-Europa* (setting aside the marginal regions) both the Moho and the basement surface have a subdued relief. The basement descends with a remarkably even gradient from the edge of the Baltic shield to depths of almost 3 km beneath the broad ENE

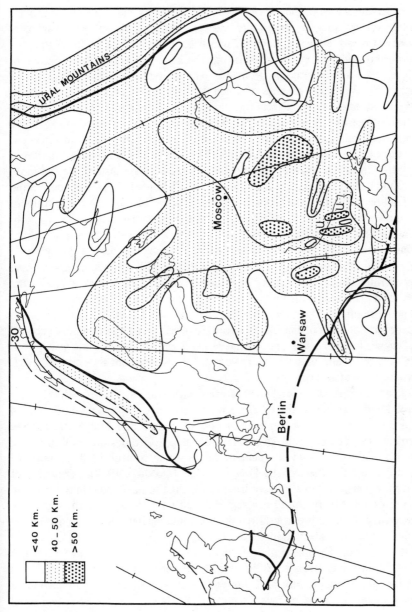

Fig. 3. Sketch contour map of the Moho beneath *Eo-Europa* based on Belyaevsky *et al.* (1973), Sellevoll (1973) and Sollogub *et al.* (1973). U = Ukraine–Donetz structures discussed here

Legend:
<40 Km.
40 – 50 Km.
>50 Km.

downwarp of the Moscow basin or central Russian depression. At the same time, the crust thickens from 35–40 km to 40–45 km but shows no very abrupt variations.

In the southern part of the craton, both the basement surface and the Moho are characterized by stronger relief and steeper gradients, though no exact correlations between them are apparent. The variations in level of the basement surface are expressed by the alternation of narrow, often fault-bounded troughs containing up to 4 km of cratonic sediments with a number of blocks such as the Ukrainian and Voronezh massifs in which the basement is at or near the surface. The thickness of the crust reflects this arrangement to some extent: for example, the crust apparently thins beneath the deep Donetz basin. Detailed studies have also revealed remarkable variations of crustal thickness which seem independent of the thickness of cover. In the Ukraine, Sollogub *et al.* (1973) have found that the depth of the Moho varies by as much as 15–20 km over distances of no more, than 100 km. Strips of thicker and thinner crust bounded by lines oblique to the great features delimiting the Donetz–Donbass basin extend from the Ukrainian massif under the basin to the Voronezh massif (Fig. 3). These strips are roughly parallel to the grain of the basement and also to the run of transverse 'uplifts' which influenced sedimentation during the Late Palaeozoic stage of infilling of the Donetz basin. The thinner strips appear to coincide with tectonic units occupied by regenerated Archaean rocks. The thicker strips, which may underlie the 'uplifts' of the Palaeozoic trough coincide with units of Early Proterozoic rocks; some show multiple reflectors at depths of about 60 km which are thought to represent old positions of the Moho. From these remarkable results Sollogub and his colleagues infer that the relief of the Moho is essentially the same as it was in Proterozoic times and that the thicker crustal strips represent the roots of Early Proterozoic mountain-belts.

Such an interpretation implies that even within areas no more than a few hundred kilometres across there has been little effective redistribution of material near the level of the Moho over a period of some 1800 Ma. Some relative movements between adjacent strips in Late Palaeozoic times are implied by the variations of thickness and sedimentary facies which control the distribution of oil- and gas-fields in the Donetz–Donbass basin (Dolenko, 1972), but the extent of these movements seems to have been too limited to have materially increased the relief of the Moho. Both on a regional and on a local scale, therefore, the evidence suggests that the structure of the upper mantle has not been substantially modified since the stabilization of the eocraton.

In this context, the time-spans of some patterns of crustal behaviour which may reflect mantle controls are of interest. There seems little doubt that the Baltic shield has acted consistently as a buoyant unit, subject at most to limited or short-term intervals of subsidence, for over 1500 Ma (Fig. 4). The relatively small thickness of the crust in this region has been ascribed by some authors (for example Sollogub *et al.*, 1973) to the effects of long-lasting uplift and erosion. There is, however, a good deal of evidence that the basement section exposed today is in most places little deeper than that which had already been exposed within a few hundred million years of the end of the Karelide cycle. Since the accumulation of

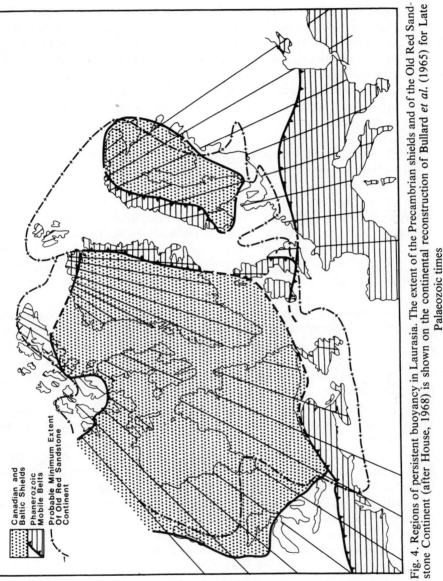

Fig. 4. Regions of persistent buoyancy in Laurasia. The extent of the Precambrian shields and of the Old Red Sandstone Continent (after House, 1968) is shown on the continental reconstruction of Bullard *et al.* (1965) for Late Palaeozoic times

Canadian and Baltic Shields

Phanerozoic Mobile Belts

Probable Minimum Extent Of Old Red Sandstone Continent

Janet Watson

the Jotnian formation, the shield appears to have been unsinkable rather than actively rising (cf. Watson, 1975). The distinctions between the Baltic shield, the central Russian depression and the more varied southernmost portion of the eocraton were established well before the end of Precambrian times, but the behaviour of the two more southerly regions did not remain entirely consistent. The central Russian depression subsided intermittently by amounts of up to 3 km over the period from Late Proterozoic to Early Mesozoic times (that is, for at least 500 Ma) but has, on balance, changed level by no more than a few hundred metres since Triassic times. The blocky structure of the southernmost region was outlined in Proterozoic times and the infilling of certain fault-bounded troughs took place in Late Proterozoic and Palaeozoic times. Its subsequent history showed considerable variations related to the development of the mobile tracts bordering the region.

The presence of deep dislocations in the craton is revealed by both geological and geophysical evidence and their importance in the architecture of the crust is manifest. Dislocations of comparatively short horizontal extent often parallel the grain of the basement structures and can be seen to be intimately related to Proterozoic tectonic patterns. Though many of these dislocations, such as those associated with the variations of crustal thickness in the Ukraine mentioned above, evidently penetrate to the mantle and have been long-lived, they may be regarded as second-order features in relation to the evolution of the craton as a whole. In contrast to them are a few persistent and long-lived lineaments which can be traced individually for distances of about 1000 km and which collectively form a set traversing almost the whole eocraton with a consistent northwesterly trend (Fig. 2). These structures appear to be largely independent of local tectonic patterns and merit individual consideration.

(a) The Karelide lineament which marks the junction of Karelide and Svecofennide subprovinces in the Baltic shield has been in existence for at least 2000 Ma. It was subject to transcurrent motions in Early Proterozoic times and may have originated as a transform structure related to the opening or closing of a small ocean basin.

(b) The Timan lineament forms the western edge of a province characterized by Late Precambrian (Baikalian) structural patterns. It is marked by conspicuous gravity and magnetic anomalies and by narrow outcrops of folded and metamorphosed Late Precambrian rocks. The lineament appears to be oblique to the grain of Early Proterozoic structures beneath the adjacent part of the Russian platform. The somewhat confusing isotopic evidence (summarized by Siedlecka, 1975) suggests that it was defined well before 1200 Ma.

(c) The Ryazan–Saratov lineament on the northeastern side of the Voronezh massif is marked by a narrow intracratonic trough containing well over a kilometre of Late Proterozoic sediments. This trough appears (Keller and Predtechensky, 1968) to have developed at or before about 1000 Ma as a rift-like structure traversing a positive block of crystalline basement. The nature of the lineament in the underlying basement and whether it originated prior to the formation of this rift are unknown.

(d) The southwestern margin of the Donetz–Donbass basin coincides with the inferred position of a Mid-Proterozoic (?1700–1500 Ma) tectonic feature attributed by Semenenko *et al.* (1968) to the effects of Volynian orogenic activity. This feature runs through an Early Proterozoic province and is oblique to the tectonic grain defined by banded iron-formations in both the Krivoi Rog and Kursk areas. Its relationships suggest that it may be a deep shear-zone on which the Donetz–Donbass graben developed in Palaeozoic times.

(e) The Tornquist line is well known through its effect on the development of Phanerozoic basins in the western part of the eocraton but comparatively little attention seems to have been paid to its early history. In Poland it coincides with the effective border of the Hercynian province and was clearly active in Palaeozoic times. Towards the northwest it leaves the Hercynian front and traverses a region underlain by an older basement. It is oblique both to the grain of the Sveconorwegian province on its northeastern side and to the inferred grain of the hidden basement on its other side. Along the southwest coat of Norway, however, a major diversion of strike brings the basement structure into parallelism with the projection of the Tornquist line in the Skagerrak. Falkum (1972) pointed out that this diversion (which he attributed to large-scale open folding) took place before the intrusion of an igneous body dated at around 1000 Ma. The Tornquist lineament, clearly identifiable up to this point (Sellevoll and Aalstad, 1971; Åm, 1975; Baartman and Christensen, 1975), does not appear to continue far into the North Sea. Summary maps such as those of Kent (1975) and Ziegler (1975) do not depict any structural element in the northern North Sea which could be attributed to the influence of this northwesterly lineament. The fact that the Tornquist line apparently stops short near the projection of the Caledonian front in the North Sea and its relationships with the Sveconorwegian basement structures suggest that the lineament originated in Precambrian times at or before 1000 Ma.

The evidence surveyed above shows that the eocraton is traversed by several large northwesterly dislocations which at their time of origin were independent of and often oblique to the local structural patterns. Most of these lineaments were in existence by 1000 Ma and one (the Karelide lineament) dates back to at least 2000 Ma. Dislocations of the northwesterly set extend in places to the present border of the eocraton and appear to have terminated at or been truncated by the structural features defining the Caledonian, Hercynian and Uralide mobile tracts. Isolated extensions of the dislocations might be sought in basement complexes within these tracts and in this connection the presence of ?-Precambrian northwesterly zones of disturbance deflecting the basement grain of the Moldanubicum in the Bohemian massif is of interest.

The length, straightness and parallel alignment of the northwesterly lineaments suggest that they may have been initiated in relation to a tectonic regime operating on a very large scale. When plotted on a globe they fall close to small circles about a pole in the vicinity of the Sea of Okhotsk. Their arrangement would be consistent with an origin as transform dislocations developed as accommodation structures during the bodily drift of a large continental mass; this possibility implies drift on lines roughly parallel to the lineaments and could therefore be tested against palaeomagnetic evidence (cf. Sutton and Watson, 1974).

The Cratonic Stage of Evolution

From the considerations outlined above, the embryonic craton formed at the end of the Karelide cycle of mobility emerges as a stable unit of at least the same order of magnitude as the present eocraton and very possibly of considerably larger dimensions. Although the affinities of the basement complexes now contained in the peripheral Palaeozoic mobile belts remain to be established, the balance of evidence suggests that most of the Scandinavian Caledonides, and possibly much of the Hercynides of central Europe (Krebs, this volume) incorporate former portions of the Mid-Proterozoic forerunner of *Eo-Europa*.

In following up the history of *Eo-Europa* as a cratonic unit, one becomes increasingly aware of the interplay of global and cratonic influences. The external relations of this unit have been radically changed by successive phases dominated on the one hand by processes of extension and disruption of the continental mass incorporating the eocraton and, on the other hand, by processes of collision and orogeny (Table 1). During these changes the internal regimes at crustal and subcrustal levels have remained in some resects remarkably consistent.

The first evolutionary stage, during which *Eo-Europa* was probably part of a

TABLE 1. *Eo-Europa* in relation to successive global tectonic systems

---

A (>1800 Ma): the pre-cratonic stage. Eo-Europa occupied mainly by mobile tectonic provinces: Archaean tectonic systems succeeded by Early Proterozoic tectonic systems in which mobile (Karelide and Svecofennide) belts surrounded many small stable massifs

B (<1800 Ma): the cratonic stage. Eo-Europa forms a single cratonic mass
   (i) (~1800–1100 Ma): the nucleus of *Eo-Europa* probably forms part of a craton of supercontinental dimensions including the Hebridean craton and parts of Greenland and North America: Sveconorwegian province added to eocraton at about the end of this phase
   (ii) (~1100–800 Ma): phase of extension and/or disruption. Positions of future Caledonides, Hercynides. Uralides determined, present limits of *Eo-Europa* defined: Hebridean craton now established as part of a separate cratonic unit
   (iii) (~800—300 Ma): complex phases of collision and orogeny in the marginal mobile belts. By the end of these phases *Eo-Europa* once more forms part of a craton of super-continental dimensions including the Hebridean craton and much of Greenland, North America, Siberia
   (iv) (~300–200 Ma): life-span of largest supercontinental craton
   (v) (~200–0 Ma): phase of extension and disruption. Opening of North Atlantic disrupts supercontinental craton, *Eo-Europa* now becomes part of Eurasian craton

---

larger craton, opened with phases of erosion revealing metamorphic and granite rocks in many Early Proterozoic provinces. The whereabouts of the erosion-products formed during these phases are uncertain, since no thick formations with the time-relationships of molasse have survived. The Jotnian sandstones, the oldest undisturbed cover-units, date from 200 Ma after the end of the Karelide cycle and are, at least in the Baltic and Ukrainian areas, neither thick nor extensive. More bulky Mid-Proterozoic formations may be concealed in the deep parts of the cratonic cover: up to 20 per cent of the total thickness in the central Russian depression appears to be Precambrian.

Records of igneous activity during the initial stages of the cratonic regime are provided both by the rapakivi–anorthosite suite, which marks a unique episode, and by occasional basic and alkaline intrusives of types which were also emplaced at later periods. The oldest alkaline complexes of the Kola peninsula, not far distant from the Palaeozoic Khibine Tundra complex, are dated at about 1700 Ma. Some alkaline and gabbroic intrusives are located in and near deep lineaments such as that at the border of the Donetz trough.

The first clearly discernible cycle involving disruption and reassembly of continental masses was that during which the Late Proterozoic–Palaeozoic mobile belts making the Caledonides, Hercynides and Uralides were developed. The present boundaries of *Eo-Europa* were blocked out initially by the fracture systems associated with the opening of graben and/or oceans on the site of these belts and were finally fixed at the orogenic fronts marking the limits of strong tectonic activity (Fig. 1). Some regional variations along these boundaries are apparent. The northwestern and eastern boundaries against the Scandinavian Caledonides and the Uralides are simple and clear-cut. The western and southern boundaries are less well defined, witness the unresolved debate concerning the extent of Caledonian activity between central England and central Europe. The variations in thickness, facies and states of deformation and metamorphism recorded from Late Proterozoic and Early Palaeozoic subcrops beneath eastern England and adjacent parts of the North Sea and the Low Countries have been held by some to relate to the presence of a concealed branch of the Caledonian orogenic system (see Kvale, this volume). The tectonic framework, however, appears to be dominated by an irregular block and basin structure not unlike that which dominated the same region in Mesozoic times and related to Cadomian or Proterozoic lines. The view that this structure was essentially cratogenic receives support from heat-flow measurements in the North Sea. Data from many boreholes suggest that, although heat-flow is high over the entire North Sea basin, there is a significant falling-off, not related to thickness of cover, from the region north of a line connecting the known Caledonian provinces of Britain and Scandinavia towards the southern North Sea. The geothermal gradients calculated by Evans and Coleman (1973) are 35–40°C/km north of this line but seldom above 35°C/km south of it.

Extension and fracturing roughly coeval with the initial stages of break-up of the Proterozoic supercontinent are recorded in the interior of the eocraton by a number of graben filled with late Proterozoic sediments. The Ryazan–Saratov trough (Fig. 1) is one of a number of structures with northwesterly or north-easterly elongation. Neither within nor at the margin of *Eo-Europa* is there evidence that large-scale cratonic magmatism accompanied the development of extensional fracture-systems—Europe has no plateau basalts or dolerites to match the Keweenawan lavas and Mackenzie dyke swarm formed at the corresponding stage in North America and, perhaps as a consequence, lacks major Proterozoic copper-deposits.

Towards the end of the Precambrian era, disturbances in the peripheral mobile provinces (Baikalian, Assyntian, Cadomian and early Caledonian episodes) led to the elevation of highlands on the borders of the eocraton and thereby established a pattern of sediment supply which was to persist with variations until the present day. From this stage onward, the main sources of detritus lay outside the craton and the main basins of subsidence were close to its borders. In the interior regions the total thicknesses of the cratonic cover accumulated over more than 500 Ma do not exceed 3 km. Long-term fluctuations in total thickness and in the relative proportion of psammitic, pelitic and chemical-organic sediment reveal a subdued response to changes in the distant sourcelands. In the marginal basins the response was more direct. Rapid subsidence is illustrated by the occurrence of up to 12 km of mainly Late Palaeozoic sediments in the Donbass basin and of thicknesses of several kilometres accumulated over relatively short periods in troughs adjacent to the Urals and Hercynides and in the North Sea. Relative vertical motions led to the juxtaposition of such basins with positive blocks in which the basement remained close to the surface, providing the environments in which the principal oilfields of the eocraton are located.

Despite the obvious importance of the peripheral highland tracts to the evolution of the eocraton through Phanerozoic times, the long-term influence of the cratonic regimes was still apparent. On a regional scale, the persistent buoyancy of the Baltic shield was sufficient to inhibit the accumulation of thick sedimentary successions even during periods when abundant detritus was available. The great ovoid uplift which constituted the Old Red Sandstone Continent formed at the close of the Caledonian orogenic cycle almost coincides with the Baltic, Greenlandic and Canadian shields. The only portions of the Palaeozoic belts which are not flanked by clastic wedges or fault-troughs containing late orogenic sediments are those adjacent to the Baltic and Greenlandic shields which had and still have a history of buoyancy (Fig. 4).

On a somewhat smaller scale, many successor basins receiving sediment from the peripheral sourcelands are seen to be set obliquely relative to the Palaeozoic orogenic fronts. The axial direction of most of these basins is parallel to older lineaments in the eocraton itself. The influence of pre-existing architectural features is well illustrated in the development of the Donbass basin which diverges obliquely from the Hercynian front, following old dislocations between the Ukrainian and Voronezh massifs. The Late Palaeozoic 'fill' of the basin, generally speaking, thins northwestward with distance from the peripheral sourceland; but smaller-scale variations, controlling the distribution of oil and gas, are related to transverse 'uplifts' which appear to correspond to the ancient strips of thick crust already referred to (cf. Sollogub *et al.*, 1973, Fig. 9, with Dolenko, 1972, Fig. 2). Narrow Palaeozoic basins of the Kamsk–Kinelsk oil and gas province are similarly oblique to the orogenic front of the Urals, several of the principal units having a northwesterly trend (Ovanesov *et al.*, 1972).

The second important phase of continental disruption affecting *Eo-Europa* was that leading to the break-up of Laurasia and the opening of the North Atlantic (Table 1). The repercussions of this phase in the main part of the eocraton were slight. No new rift-structures were formed east of the Tornquist line and scarcely any cratonic igneous activity took place in this region: there are no equivalents of the Siberian traps or Karroo dolerites emplaced during the corresponding period of disruption in Siberia and Africa. Kimberlites are also almost lacking, though the presence of detrital diamonds in the Timan range (Metallogenic Map of Europe) suggests that kimberlitic bodies analogous to those near the Nagssugtoqidian front (Escher and Watterson, 1973) may have invaded the Timan lineament.

In western Europe, rifting and block-faulting in Late Palaeozoic and Early Mesozoic times affected a crescentic tract extending from the northern North Sea through Britain and into central Europe. This tract belongs mainly to the Caledonian and Hercynian tectonic provinces but also includes the 'Danish triangle' between the Tornquist line and central England, which appears to be floored by Sveconorwegian and Cadomian basement complexes, as well as the Hebridean craton which is floored by a much older basement (the Lewisian complex, stabilized around 1800 Ma). Through the entire tract, so far as is known, the crust averages little more than 30 km in thickness as against the thicknesses of 35–45 km characteristic of most of the eocraton. Igneous activity in Permian, Mesozoic and Tertiary times led to the formation of a variety of pipes, plugs and central complexes on and near dislocations and to more extensive lava-plateaux both in the eocraton (the Oslo graben and part of the Hebridean province) and in the Palaeozoic provinces. Heat-flow data from the North Sea (Evans and Coleman, 1973) suggest that geothermal gradients are somewhat steeper than those in the interior part of the eocraton where the cratonic cover is of comparable thickness.

The spatial coincidence of the area of thin crust with that penetrated by ramifying fractures and containing deep Mesozoic–Tertiary sedimentary basins, together with the parallel evolution of the ensialic North Sea basins and those at the newly developed Atlantic margin suggest that all these features originated in response to the extensional regime that led ultimately to the opening of the Atlantic. Many of the controlling fractures appear, however, to be related to much older dislocations. The Oslo graben, for example, is a Permian structure: but the earliest alkaline igneous centre spatially related to the feature (the Fen complex) is Early Palaeozoic and the western rift-fault is parallel to and little removed from a basement dislocation (the 'great friction breccia') at least 1000 Ma in age. At the Atlantic margin of northwest Britain, the fault-bounded marginal troughs which cross the continental shelf (Bott and Watts, 1970) and separate the Outer Hebrides from mainland Scotland are essentially parallel to the (fault-controlled?) margins of Torridonian basins of deposition which date from the Late

Proterozoic phase of continental disruption and which are, in turn, parallel to early dislocations in the Lewisian basement.

The system of northwesterly fault-troughs and positive blocks (the *Randtroge* discussed by Voigt, 1963) which form the principal Mesozoic structures of the southern North Sea and adjacent parts of Denmark, Germany and the Netherlands is bounded on the east by the Precambrian Tornquist line and is paralleled on the west by linear magnetic anomalies defined by pre-Palaeozoic rocks or structures in central England. These linear troughs stop short south of known outcrops of the Caledonian province and may be aligned parallel to pre-Caledonian basement lineaments of the same set as the Tornquist line; the central graben of the North Sea itself abruptly changes direction near the inferred position of the Caledonian front. Towards the south, linear Mesozoic basins of north-westerly trend flank some of the Hercynian massifs of central Europe. Ancient (Assyntian or pre-Assyntian) structures appear to be preserved with little modification in some of these massifs and if, as Krebs suggests (this volume), Hercynian structural development was controlled largely by vertical movements, it seems possible that pre-Hercynian lineaments originally related to those traversing *Eo-Europa* may have continued to influence the evolution of orogenic and post-orogenic structures.

The most conspicuous effects of Mesozoic block-movement and graben-formation related to the opening of the Atlantic were concentrated within the crescentic region which had been characterized by fracturing and block movements during Late Proterozoic times. The mantle-generated stresses of Mesozoic times may, therefore, have found this part of the crust already segmented by deep fractures on which movements were renewed in response to these stresses. The apparent attenuation of crust in the Danish triangle relative to that of the main eocraton may itself represent a response not only to the Phanerozoic but also to the Proterozoic phase of continental extension.

Crust–Mantle Relationships

A remarkable consistency of behaviour is revealed by this very general review of the evolution of the *Eo-European* craton over 1800 Ma. This consistency suggests, on the one hand, that the craton is and has been reasonably well adjusted to the tectonic and thermal regimes at depth and, on the other, that these regimes have themselves changed comparatively little. In particular, the paucity of igneous activity over the past 1200 Ma and the persistent low topographical relief suggest that crustal and upper mantle temperatures did not fluctuate widely with time.

If the results of deep seismic studies discussed above are accepted as indications that the relief of the Moho has remained more or less the same since Early Proterozoic times, it seems to follow that the regional variations in density necessary for the maintenance of isostatic equilibrium were also established soon after the stabilization of the eocraton. Unless these variations are confined to the lower crust—a possibility rendered unlikely by the broad similarities of upper crustal structure over very wide areas—it would also follow that the architecture of the upper mantle was fixed in outline in Proterozoic times. Taken together these inferences are consistent with a picture of the eocraton as the upper part of a thick slab of ancient lithosphere which has not been substantially modified either by tectonic processes or by thermal changes associated with the development of mantle upcurrents or the over-running of hot spots. This lithospheric slab has moved bodily since its formation over distances of many thousands of kilometres without losing its identity or changing its internal regimes.

References

Ager, D. V. (1975). 'The geological evolution of Europe', *Proc. Geol. Ass.*, **86**, 127–54

Åm, K. (1975). 'Geophysical indications of Permian and Tertiary igneous activity in the Skagerrak', *Norsk. Geol. Unders. Bull.*, **13**, 1–25

Baartman, J. C. and Christensen, C. B. (1975). 'Contributions to the interpretation of the Fennoscandian border zone', *Danmarks Geol. Unders. II Ser.*, 102

Belyaevsky, N. A., Borisov, A. A., Fedynsky, V. V., Fotiadi, E. E., Subbotin, S. I. and Volvosky, I. S. (1973). 'Structure of the earth's crust on the territory of the U.S.S.R.', *Tectonophysics*, **20**, 35–46

Bott, M. H. P. and Watts, A. B. (1970). 'Deep sedimentary basins proved in the Shetland–Hebridean continental shelf and margin', *Nature, Lond.*, **225**, 265–8

Bridgwater, D. and Windley, B. F. (1973). 'Anorthosites, post-orogenic granites, acid volcanic rocks and crustal development in the North Atlantic shield during the mid-Proterozoic', *Geol. Soc. S. Africa, Sp. Pub.*, **3**, 307–18

Bullard, E. C., Everett, J. E. and Smith, A. G. (1965). 'The fit of the continents around the Atlantic', *Phil. Trans. Roy. Soc. Lond.*, **A,258**, 41–51

Dolenko, G. N. (1972). 'Regularities of oil and gas accumulations on the territory of the Ukrainian S.S.R.', *Int. Geol. Congr. 24th Session*, **5**, 75–81

Escher, A. and Watterson, J. (1973). 'Kimberlites and associated rocks in the Holsteinsborg–Søndre Strømfiord area, central West Greenland', *Rapp. Grønlands Geol. Unders.*

Evans, T. R. and Coleman, N. C. (1973). 'North Sea geothermal gradients', *Nature, Lond.*, **247**, 28–30

Falkum, T. (1972). 'On large-scale tectonic structures in the Agder–Rogaland region, southern Norway', *Norsk Geol. Tidssk.*, **52**, 371–6

Gaal, G. (1972). 'Tectonic control of some Ni–Cu deposits in Finland', *Int. Geol. Congr. 24th Session*, **4**, 215–24

House, M. R. (1968). 'Continental drift and the Devonian system', *Inaugural Lecture*, University of Hull

Kahma, A. (1973). 'The main metallogenic features of Finland', *Geol. Surv. Finland, Bull.* **265**

Keller, V. M. and Predtechensky, N. N. (eds.) (1968). Palaeogeographical and palaeotectonic maps of the U.S.S.R., Moscow

Kent, P. E. (1975). 'Review of North Sea basin development', *Journ. Geol. Soc. Lond.*, **131**, 435–68

Khain, V. E. (1976). 'The new international tectonic map of Europe' (this volume)

Kratz, K. O., Gerling, E. K. and Lobach-Zhuchenko, S. B. (1968). 'The isotope geology of the Precambrian of the Baltic Shield', *Can. Journ. Earth Sci.*, **5**, 657–60

Kratz, K. O., Gerling, E. K. and Lobach-Zhuchenko, S. B. (1974). 'Geological and geochronological boundaries in the Precambrian of the Baltic Shield', *Précambrien des zones mobiles de l'Europe*, Prague, 277–82

Krebs, W. (1976). *'Meso-Europa'* (this volume)

Nalivkin, D. V. (tr. Rast, N.) (1970). *Geology of the U.S.S.R.*, Oliver and Boyd

Ovanesov, G. P., Abrikosov, I. Kn and Khachatryan, R. O. (1972). 'Reefs of Kamsk–Kinelsk system of troughs and their role in the process of oil accumulation', *Int. Geol. Congr. 24th Session*, 5, 99–103

Piper, J. D. A. (1975). 'Palaeomagnetic evidence for a Proterozoic supercontinent', *Phil. Trans. Roy. Soc. Lond.*, A,280, 469–90

Rickard, D. T. and Zweifel, H. (1975). 'Genesis of Precambrian sulfide ores, Skellefte district, Sweden', *Econ. Geol.*, 70, 255–74

Sederholm, J. J. (1930). 'The Pre-Quaternary rocks of Finland', *Bull. Comm. géol. Finlande*, 160

Sellevoll, M. A. (1973). 'Mohorovicic discontinuity beneath Fennoscandia and adjacent parts of the Norwegian Sea and North Sea', *Tectonophysics*, 20, 359–66

Sellevoll, M. A. and Aalstad, I. (1971). 'Magnetic measurements and seismic profiling in the Skagerrak', *Mar. Geophys. Res.*, I, 284–302

Semenenko, N. P., Scherbak, A. P., Vinogradov, A. P., Tougarinov, A. I., Eliseeva, G. D., Cotlovskay, F. I. and Demidenko, S. G. (1968). 'Geochronology of the Ukrainian Precambrian,' *Can. Journ. Earth Sci.*, 5, 661–72

Siedlecka, A. (1975). 'Late Precambrian stratigraphy and structure of the northeastern margin of the Fennoscandian Shield (East Finnmark–Timan region), *Norges Geol. Unders. Bull.*, 29, 313–48

Sollogub, V. B., Litvinenko, I. V., Chekunov, A. V., Ankudinov, S. A., Ivanov, A. A., Kalyzhnaya, L. T., Kokorina, L. K. and Tripolsky, A. A. (1973). 'New D.S.S.-data on the crustal structure of the Baltic and Ukrainian Shields', *Tectonophysics*, 20, 67–84

Sutton, J. and Watson, J. (1974). 'Tectonic evolution of continents in early Proterozoic times', *Nature, London*, 237, 433–5

Sveriges Geologiska Undersökning (1967). Aeromagnetic maps of Kiruna. *S.G.U. Ser. Af.* Nr 1–4

Voigt, E. (1963). 'Über Randtroge vor Schollernrändern und ihre Bedeutung im Gebiet der Mitteleuropäischen Senke und angrenzender Gebiete', *Z. dt. Geol. Ges.*, 114, 378–418

Watson, J. (1975). 'Vertical movement in Proterozoic tectonic provinces', *Phil. Trans. Roy. Soc. Lond.* (in press)

Ziegler, P. (1975). 'The geological evolution of the North Sea area in the tectonic framework of northwestern Britain', *Norges Geol. Unders., Bull.*, 29, 1–28

Palaeo-Europa

That part of the continent unaffected by major
tectonism since the end of Early Palaeozoic times

MAJOR FEATURES OF THE EUROPEAN
CALEDONIDES AND THEIR DEVELOPMENT

ANDERS KVALE

*Geologisk institutt, Avdeling A, Universitetet
i Bergen, N-5014 Bergen, Norway*

Abstract

Kvale, A. (1976). 'Major Features of the European Caledonides and their Development', in Ager, D. V. and Brooks, M. (eds.), *Europe from Crust to Core*, Wiley, London. The limits of the Caledonian orogeny in time and space as well as the extent of the Caledonian geosynclinal area in Europe are briefly discussed, Greenland being assumed to have had the pre-Tertiary position as given by Bullard *et al.* (1965). Sedimentation and volcanic activity in Late Precambrian and Eocambrian time are reviewed. The main features of sedimentation and volcanic activity in the Caledonian geosynclinal area in Europe in Cambrian to Silurian times are outlined, with emphasis on a comparison between the development in the northwestern and the continental branches of the European Caledonides. The orogenic phases and events which have been supposed to occur in Europe during the Caledonian orogeny are discussed, and temporal and spatial correlations are attempted. A separation of metamorphic zones in the European Caledonides is proposed.

It is concluded that in our present state of knowledge research should be concentrated on the history of development of the European Caledonides. This would increase the reliability of applications of plate tectonic models to the European Caledonides and enhance the scientific value of such applications.

Our first task is to define the Caledonian orogeny of Europe in time and space. Among Scandinavian geologists it has been usual to consider the youngest Precambrian sedimentation, the Eocambrian, as indicating the beginning of the development of the Caledonian geosyncline, and the formation of Early and Middle Devonian deposits in intermontane basins and the movements affecting them before the Late Devonian as marking the end of the Caledonian era. As regards the limits in space, a study of the Caledonides in Europe would be incomplete if Spitzbergen and East Greenland were not included.

The Caledonian Orogen

We may define the Caledonian orogen as the area where the rocks were affected by Caledonian orogenic movements. There seems to be general agreement on the boundaries of that orogen in Scandinavia, Spitzbergen, East Greenland and Great Britain. But opinions differ widely as to the extent of the Caledonian

81

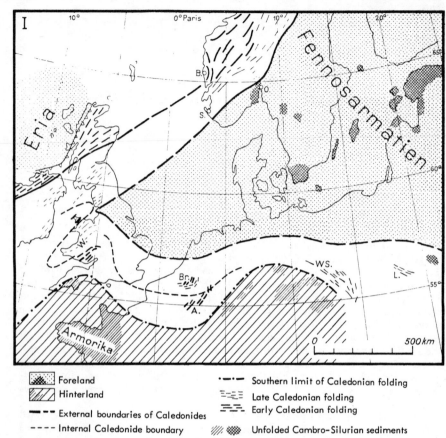

Foreland

Hinterland

External boundaries of Caledonides

Internal Caledonide boundary

Southern limit of Caledonian folding

Late Caledonian folding

Early Caledonian folding

Unfolded Cambro-Silurian sediments

Fig. 1. Possible connections of Caledonian folded areas in northern central Europe. B. Bergen; S. Stavanger; O. Oslo; W. Wales; L. Lysagora; W. West Sudeten; Br. Brabant Massif; A. Ardennes Massif. (After von Gaertner, 1960, Fig. 1)

orogen in continental Europe. Von Gaertner (1960; Fig. 1) drew the southern limit of Caledonian folding through Cornwall, south of the Ardennes and south of the West Sudeten. The present width of the orogen in this area would then be from 200–400 km. Bogdanoff *et al.* (1964; Fig. 2) included the whole of central and southern Europe as well as part of North Africa in the Caledonian orogen, thereby widening it to between 1500–2000 km. This assumption seemed to be based on the occurrence of 'Caledonian geosynclinal folded formations' in southern France, in northern and southern Spain, in Sardinia and Corsica, in Calabria and Sicily and in the border area of Turkey and Lebanon.

On the Metamorphic Map of Europe (1973) all of these areas in southern Europe were marked as having been metamorphosed during the Variscan orogeny. This map does, however (Fig. 3), include areas in the south of France and in the Alps as having been metamorphosed during the Caledonian orogeny. The present width of the Caledonian orogen in France and Belgium would then be at least 800 km. Obviously there is a need for more studies on this problem.

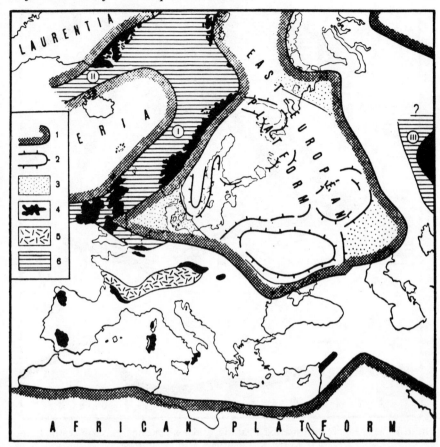

Fig. 2. Caledonian geosynclinal area and adjoining platforms: 1, presumed borders of platforms; 2, outlines of shields; 3, most deeply subsided and, apparently, faulted corners of the platforms; 4, outcrops of Caledonian geosynclinal folded formations; 5, systems of 'Franco-Podolian' folded masses, partly regenerated by Caledonian folding; 6, areas where geosynclinal development was finished by Caledonian folding. (After Bogdanoff, Mouratov and Schatsky, 1964)

The East European Platform defines the eastern limit of the northern European Caledonides and the northern limit of the central European Caledonides. The boundary of the orogen may be defined as the line beyond which the Caledonian deformation of the Precambrian rocks is absent or negligible. In Scandinavia this line can be determined with some accuracy, thanks to good exposures (Fig. 4). As might be expected the boundary is not a line, but a zone of transition, which is at most a few tens of kilometres wide. The zone does not, however, coincide with the eastern limit of the folded or thrust Cambro–Silurian rocks. The difference is most clearly seen in southern Norway, where the cover rocks were folded and thrust in an area extending up to about 200 km above the undeformed Precambrian basement (Fig. 4).

The western border of this part of the Caledonian orogen is partly hidden

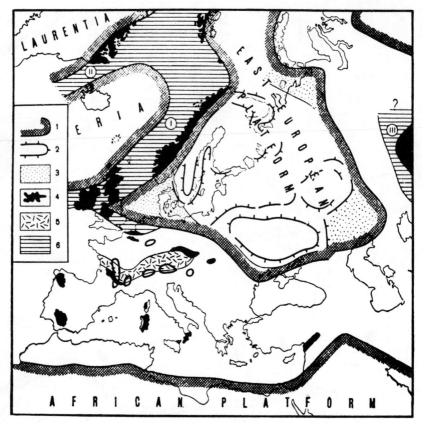

Fig. 3. Same as Fig. 2, with inserted areas of Caledonian metamorphism in central Europe
according to the Metamorphic Map of Europe 1973

beneath the ice of East Greenland (Fig. 5). In spite of the meagre evidence there
seems to be fairly good agreement regarding its approximate position. The line on
this map is drawn after Haller (1970). Greenland is here given the pre-Tertiary
position assumed by Bullard *et al.* (1960). With Greenland in this position the
western border of the orogen lines up fairly well with the northwestern border in
the north of Scotland.

The southern limit of the East European Platform and especially its westward
extent has until recently not been well known, as can be seen from Fig. 6. Some
authors have favoured a direct connection between the Caledonides in Scandinavia
and those in central Europe, while others have assumed that the southwestern corner of the platform lies under the North Sea or in Great Britain. According to
Sorgenfrei and Buch (1964) the drilling for oil in Denmark gave no indication of
Caledonides in the Danish subsurface. Geophysical investigations and drill cores
from the northern part of central Europe have in recent years confirmed the westward extension of the East European Platform in the North Sea (Pozaryski *et al.*,
1965). According to Kent (1967), aeromagnetic evidence of moderately shallow

WESTERN BOUNDARY OF
UNDEFORMED PRECAMBRIAN
BASEMENT

EASTERN BOUNDARY OF
FOLDED OR THRUST
CAMBRO-SILURIAN

Fig. 4. Boundaries of Caledonides in Scandinavia

basement in the neighbourhood of the Humber mouth and off Norfolk supports the idea of a westward extension of the platform to England.

The southern border of the East European Platform should therefore now be fairly well known. Whether it coincides with the northern border of the Caledonian orogen in central Europe is more difficult to determine. The northernmost outcrops of Caledonian fold belts lie between 100–200 km south of the platform border, while in Scandinavia the fold belt extends 200 km above the platform. This difference gives evidence of the great difference in intensity of the horizontal movements in the two branches of the European Caledonides.

The southern limit of the Caledonian orogen in Europe is not known for two reasons: (i) areas of known Cambro–Silurian rocks are few and far apart (Fig. 7) and (ii) Caledonian structures may be overprinted by Variscan or Alpine structures or by both. The limit certainly must lie south of the Caledonian fold belts in the south of France and in the Alps. It must also lie north of the northern boundary of the African Platform as given by Bogdanoff *et al.* (1964). On this problem also, there is a need for further research.

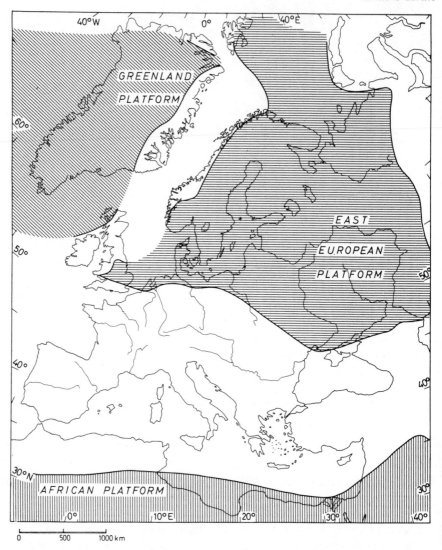

Fig. 5. Europe with the Greenland Platform in assumed pre-Tertiary position

The Caledonian Geosynclinal Area

It has been questioned whether the Caledonian structures through Middle Europe were formed in a true geosyncline (Störmer, 1967). In Britain it seems probable (Bennison and Wright, 1969) that sedimentation in the geosyncline took place in several separate basins which were not always elongate in form, and that different areas within the basins received their maximum sedimentation at different times. In Scandinavia also it may be questioned to what extent sedimentation followed the pattern set forth in definitions of geosynclines. Part of the geosynclinal

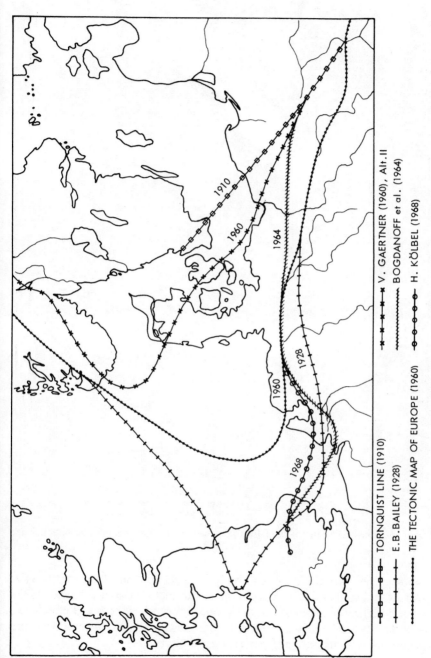

Fig. 6. Assumed borders of East European platform. (After Störmer, 1967, Fig. 5)

TORNQUIST LINE (1910)

E.B.BAILEY (1928)

THE TECTONIC MAP OF EUROPE (1960)

V. GAERTNER (1960), Alt.II

BOGDANOFF et al. (1964)

H. KÖLBEL (1968)

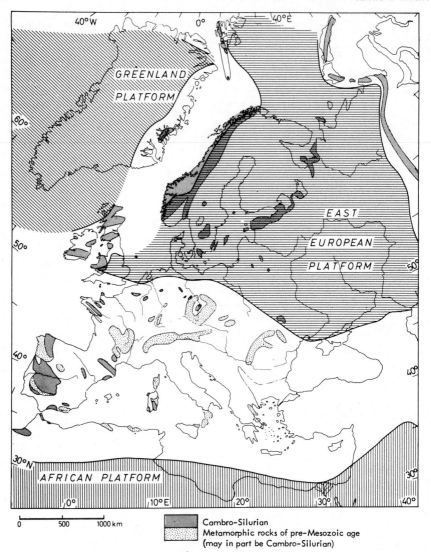

Fig. 7. Cambro–Silurian rocks in Europe. (After Carte Géologique international de
l'Europe)

terminology has, however, proved useful for a description of the stratigraphic
sequences in Scandinavia, as was shown by Störmer (1967) and Strand (1972).
Störmer described the lithology of the various geosynclinal belts and presented a
map suggesting their primary position. In Fig. 8 his zones are transferred to a map
with Greenland in its assumed pre-Tertiary position.

Going from east to west we find the deposits of the craton, the foreland, the
miogeosynclinal and two eugeosynclinal belts with different stratigraphy. Strand
showed the present distribution of the various sequences on a map which is, with

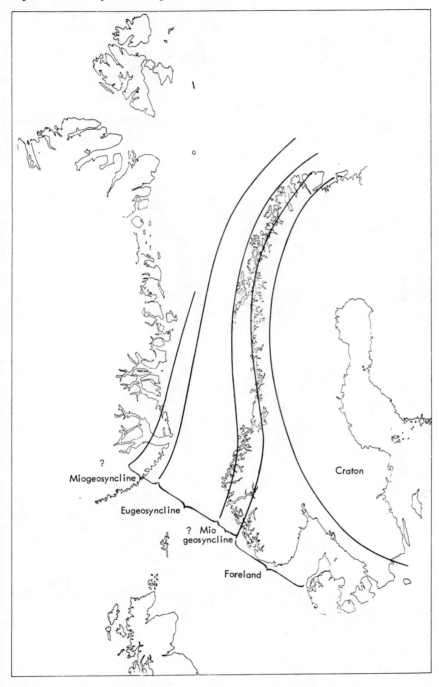

Fig. 8. Geosynclinal belts in Scandinavia. (After Störmer, 1967, Fig. 5) transferred to map with Greenland in assumed pre-Tertiary position

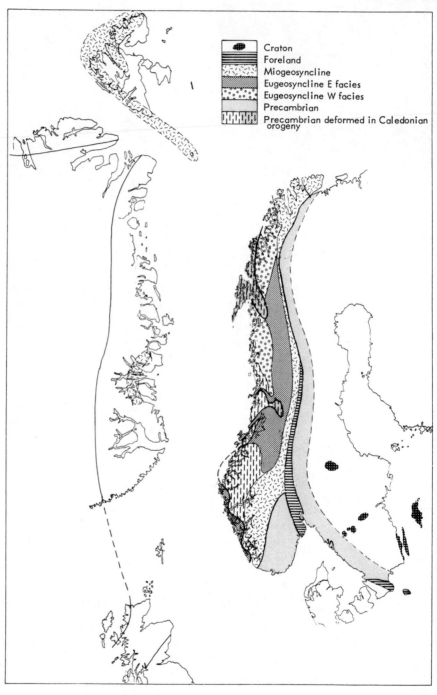

Craton
Foreland
Miogeosyncline
Eugeosyncline E facies
Eugeosyncline W facies
Precambrian
Precambrian deformed in Caledonian orogeny

Fig. 9. Present distribution of geosynclinal facies in Scandinavia, Spitzbergen and East Greenland. (Partly after Störmer (1967) and Strand (1972))

some modifications, used in Fig. 9. No evidence of a western miogeosynclinal belt has been found in Scandinavia. The western eugeosynclinal facies in northern Norway bears some resemblance to the Cambro–Ordovician of Spitzbergen and East Greenland, the latter of which Haller (1970) found difficult to characterize either as miogeosynclinal or eugosynclinal, though the rocks generally show a miogeosynclinal aspect.

The autochtonous Cambro–Ordovician in the northwest of Scotland, which resembles the East Greenland sequence in having thick series of limestone and dolomite, has been considered to have been deposited on a wide shelf area extending into North America.

We may conclude that there is no evidence of a western miogeosynclinal belt comparable to the eastern belt in Scandinavia, where argillaceous and arenaceous beds predominate.

That part of Europe which lies between the East European and the African platforms is characterized by Brinkmann (1969) as belonging to 'a wide complex geosynclinal area'. Few attempts have been made to separate geosynclinal belts. Kölbel (1968) presented a map of central Europe indicating the palaeogeography in Cambrian to Silurian times (Fig. 10). He separated one miogeosynclinal belt in the north, another south of the Alps, and between them two eugeosynclinal belts and two anticlinal ridges, the northern one being temporary.

Depositional History of the Geosynclinal Area

The latest Precambrian orogeny which affected the East European platform was the Sveconorwegian or the Dalslandian orogeny. It probably ended between 1000–1100 Ma ago, but post-tectonic granites have ages as low as 850–900 Ma. The platform has since then only been subjected to epeirogenic movements. West of its present border, rocks of the same ages were severely affected by the Caledonian orogeny, as can be studied in the coastal areas of Norway between 59 and 70°N. The Greenland platform consolidated after the Karolinidean orogeny, and the African platform also consolidated long before the Cambrian. In the areas between the three platforms the later part of the Precambrian was, however, not a stable period. Perhaps this period is rather outside the present topic but it will be useful, in order better to understand the development of the Caledonides, to review briefly what we may call the prelude to the Caledonian era: the events which occurred in the same general area before the beginning of the Cambrian.

Late Precambrian and Eocambrian Deposits

After the consolidation of the three platforms, sediments were deposited in various parts of the areas between them during a period of at least 300 Ma before the beginning of the Cambrian, or about twice as long as the Cambrian, Ordovician and Silurian periods together (Fig. 11). Little is known about the former extent and the shape of the areas of deposition. Some of the sedimentary series had a eugeosynclinal, others a miogeosynclinal, character. Volcanic activity was widespread throughout the period in the area between the East European and the

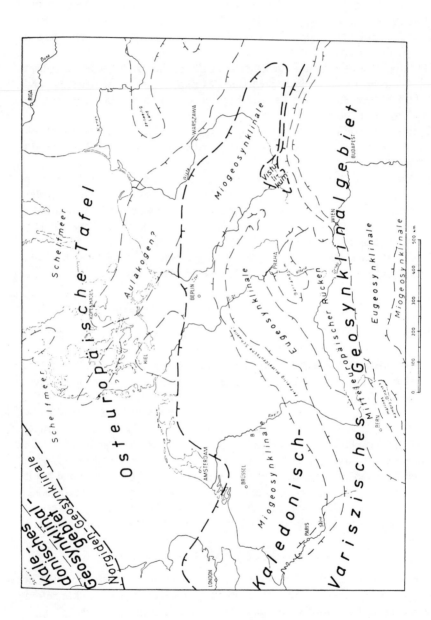

Fig. 10. Palaeogeographical sketch: Cambrian–Silurian. (After Kölbel, 1968, Fig. 10)

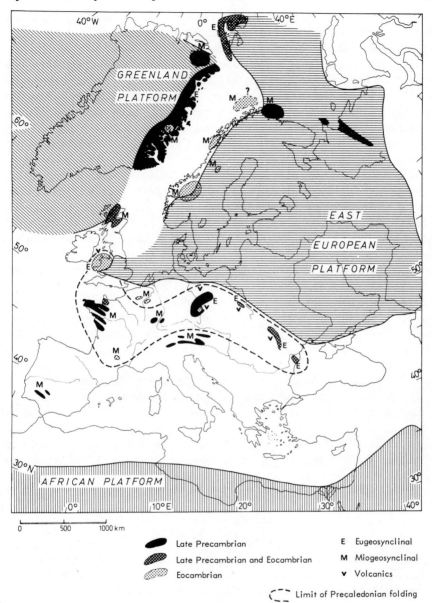

Late Precambrian

Late Precambrian and Eocambrian

Eocambrian

E   Eugeosynclinal

M   Miogeosynclinal

v   Volcanics

Limit of Precaledonian folding

Fig. 11. Late Precambrian and Eocambrian deposits in Europe

African platforms. Between the East European and the Greenland platform volcanism is recorded only early in the period, in Greenland, Spitzbergen and the Celtic Cycle area of Great Britain (Bennison and Wright, 1969, p. 63). In most of the basins the thickness of the supracrustal rocks is between 6000–9000 m. The lack of fossils and scarcity of age determinations complicate the correlation of these deposits. The older series have been called Late Precambrian or Algonkian, the younger ones Eocambrian in Scandinavia and Intracambrian in continental Europe. Other names have also been used.

The Late Precambrian and Eocambrian deposits were affected by pre-Caledonian orogenic movements in the Celtic Cycle area, in Brittany, the Massif Central in France and the greater part of central and southeastern Europe. The orogeny or the orogenic phases have been called Assyntian, Baikalian or Cadomian. In some areas the activity lasted well into the Cambrian. In the Metamorphic Map of Europe (1973) the upper time limit for these phases is given as 545 Ma or middle Cambrian.

Sedimentary series of Late Precambrian or Cambro–Silurian age are known from a few areas of the East European platform. The thickness is in most areas a few hundred metres, and the sediments may be accompanied by volcanics. Thicker series of sediments are found along the northern edge of the Kola peninsula and in east Finnmark, where the thickness is at least 9000 m (Siedlecka and Siedlecki, 1972), and in the Hagenfjord basin of northeast Greenland, where the thickness approaches 2000 m (Haller, 1970). Both of these basins lie near the border of a platform.

Orogenic activity on the East European platform is known only from the Kanin–Timan range, where 14,500 m of Late Precambrian miogeosynclinal sediments were folded on NW-trending axes, metamorphosed in greenschist facies and intruded first by gabbroic and later by granitic magma. Radiometric ages give 640–620 Ma for the intrusion of gabbros and, for a second period of metamorphism, around 525–520 Ma (Siedlecka, 1975). The figures indicate activity beginning in the Eocambrian and lasting into the Late Cambrian.

During the long interval between the consolidation of the platforms and the beginning of the Cambrian there were thus great differences, especially in volcanic activity and in orogenic activity, between the two areas which were later to become the two branches of the European Caledonides. Major differences persisted also during the Caledonian era.

Cambrian

From the Early Cambrian on, the fossil record permits a much better correlation of strata in different areas than it is possible in the older sediments. Studies of important problems such as the distributions of land and sea can also be based on more reliable evidence. But the results may still be uncertain, even in the Baltic area, where no orogenic movements or metamorphism have disturbed the record (Fig. 12). Martinson (1974) found that for each of six subdivisions of the Cambrian there exist from three to five different palaeogeographical maps (Fig. 13). The construction of palaeogeographical maps from the Caledonian geosynclinal area should be expected to be much more difficult than from the Baltic, but some of the main features seem to be fairly well established.

In Early Cambrian times the sea transgressed over large areas in Europe (Fig. 7). In the Middle and Late Cambrian the sea advanced in some areas and

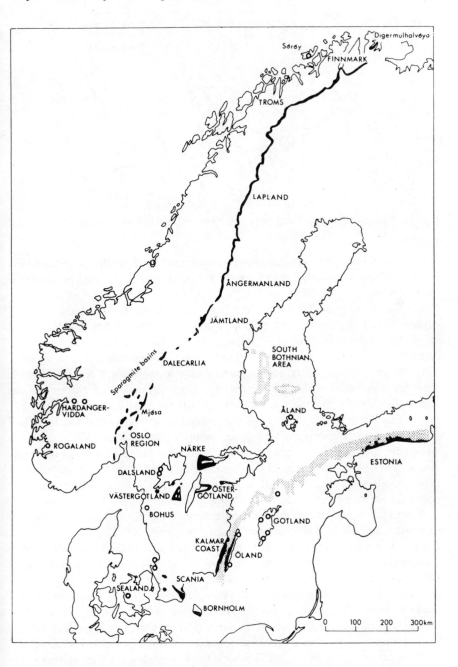

Fig. 12. Distribution of Cambrian deposits in Scandinavia. Submarine exposures are stippled. Rings mark borings or local occurrences. (After Martinson, 1974, Fig. 1)

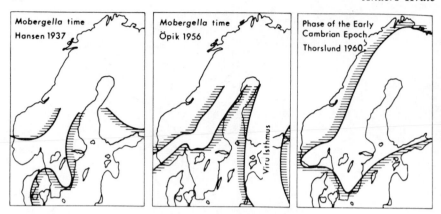

Fig. 13. Palaeogeography of the Early Cambrian in Scandinavia according to different authors. (After Martinson, 1974, Fig. 11)

retreated in others. Following the orogenic movements in Late Precambrian times basins of Cambrian deposition developed in various parts of the area between the three platforms.

The Cambrian deposits south of the East European platform generally have a eugeosynclinal character and a thickness of from 1000–3000 m in the various areas. Although the evidence is incomplete, it seems likely that the sediments were deposited in several basins or synclines rather than in one continuous geosyncline. Volcanic activity was scattered and mostly weak in the Early Cambrian, but in the Late Cambrian volcanoes were active from Brittany and Normandy to Romania, as well as in the southeastern part of the East European platform.

In Great Britain about 4500 m of eugeosynclinal sediments and volcanics accumulated in the Welsh basin during the Cambrian (Bennison and Wright, 1969). Along the southeastern margin of the basin the sequence has a miogesynclinal character (Dalziel, 1969). Autochthonous Cambrian deposits are lacking between the Midlands and the northwest of Scotland, where the Cambrian is dominated by dolomitic limestones, as in East Greenland and Spitzbergen.

Across the North Sea, in the foreland sediments of the Oslo area, a complete Cambrian section is found, having a total thickness of 120 m. The Early Cambrian is found only in the northern part of the area, but the Middle Cambrian transgression probably covered most of southern Norway. The typical rock of the Late Cambrian is an alum shale which, according to Björlykke (1974) was deposited at the rate of 1 mm in 1000 years. The alum shale is also found in the miogeosynclinal sequence at Hardangervidda in central Norway, where the thickness of the Cambrian is about the same as in the Oslo area. The thickness of the Cambrian in the eugeosynclinal sequence is not known, but alum shale phyllites occur in the lower parts. The total thickness below the Early Ordovician volcanics seems to be only a few hundred metres, at least in some areas.

It appears that the rate of sedimentation in Cambrian times was extremely slow over southern Norway and adjoining parts of the North Sea. It is hardly justified to speak of a geosynclinal area. The Cambrian greywacke in the Iltay Nappe in the Scottish Highlands does, however, indicate geosynclinal conditions in the area where these sediments originated.

In northern Scandinavia the Cambrian beds are thicker. In east Finnmark the basal Cambrian sandstones are 200 m thick as against a few metres in the Oslo area. At Sörōy in west Finnmark about 1000 m of sediments, including greywackes, contain fossils of Early and Middle Cambrian age (Sturt and Ramsay, 1965; Pringle and Sturt, 1969). This sequence indicates the existence of a Cambrian geosynclinal area north of the area of extremely slow sedimentation. Whether it should be called eugeosynclinal or miogeosynclinal is a matter for discussion, but Cambrian volcanics are lacking throughout the northern area.

Ordovician

The Ordovician of continental Europe is dominated by shales and sandstones. Thicker deposits of carbonates occur in Hungary, and the Late Cambrian volcanic activity continued in some areas into the Early Ordovician. The thickness of the supracrustal rocks is several hundred metres, in some areas between 1000–2000 m. The deposits may be characterized either as foreland or as miogeosynclinal.

The only exception to this uniform pattern seems to be the Barrandian Basin of Bohemia (Svoboda *et al.,* 1966) where, in an area of about 100 × 30 km a typical eugeosynclinal section is preserved, containing a variety of sediments, including greywackes and cherts, and showing great local variations. The sequence spans the whole of the Ordovician period. Volcanic activity was moderate, and the products were basic. The eugeosynclinal facies may have had a wider distribution, as was indicated in the map by Kölbel.

In Great Britain sedimentation continued in Wales throughout the Ordovician. North of the Irish Sea geanticline another basin developed, extending from the Lake District to southern Ireland. A third basin developed from the Southern Uplands to northwest Ireland. The present width of the deposits in each basin is from 100–150 km, and the ridges between them have a similar width. The deposits in all basins have a eugeosynclinal character, and their thickness exceeds 3000 and 4000 m in Wales and in the Lake District respectively. The volcanic activity was especially heavy in Wales, where it lasted from Late Tremadoc to Late Caradoc, beginning with basic and continuing with more acid volcanics.

To the northwest, sediments in the Girvan area of Scotland have a shelf character. To the southeast the English Midlands were not submerged.

In southern Norway the sea of the alum shale lasted through the Early Tremadocian. Then the conditions of sedimentation changed, after a stable period of perhaps 30 Ma. In the foreland or shelf deposits of the Oslo area shales, calcareous shales and limestones alternate throughout the Ordovician and the greater part of the Silurian with a few sandy beds occurring, especially in the uppermost Ordovician.

West of the foreland nearly all rocks are parautochthonous or allochthonous, and we can only determine in a very general way the areas of deposition.

In the miogeosynclinal area shales predominate, but sandstones and minor layers of calcareous sediments occur. In Norway fossils prove that sedimentation continued into Llandeilo times, but there the record is broken by a thrust plane. Traces of volcanic activity are found only in the westernmost parts of the miogeosynclinal deposits.

The eastern eugeosynclinal belt has a typical eugeosynclinal sequence with a great thickness of volcanic rocks, from basic to acid, a wide variety of sediments including greywackes and cherts, and considerable local variations in stratigraphy. They are found in three separate areas, which may have been continuous, having a total length of 1300 km, but conditions of sedimentation varied with place as well as with time. A maximum width of 150 km is found in the Trondheim area, but nothing is known about the original width.

The western eugeosynclinal belt has less volcanic rocks, more carbonate rocks and less local variations in stratigraphy. It occurs only north of the Trondheim district, in two separate areas, having a total length of about 750 km and a maximum present width of about 100 km. The basins may have been larger than in the eastern belt. Beds rich in sedimentary iron ore occur along a length of 550 km. This facies bears some resemblance to the Cambro–Ordovician of Spitzbergen and East Greenland.

In Spitzbergen only the Early and Middle Ordovician, and in East Greenland only the Middle Ordovician, are represented.

It appears that a different development in the two branches of the European Caledonides, evident in the Cambrian, continued into the Ordovician but with a change of character in both branches. In continental Europe the eugeosynclinal series of the Cambrian was succeeded by miogeosynclinal or foreland deposits during the Ordovician, the Barrandian basin being the only known exception (Svoboda *et al.*, 1966). Low relief must have prevailed, and the volcanic activity was inconspicuous.

In the northwestern branch the quietness of the Cambrian was succeeded by increased activity during most of the Ordovician. The sediments give evidence of repeated uplift and erosion and the volcanic activity was considerable, especially in the Arenig. The sources of the sediments were probably islands in the geosynclinal sea. Our present evidence is more compatible with the existence of several minor or middle-sized islands than with a few large islands or semicontinents, although much more information is needed.

The Welsh basin, which is situated where the two branches meet, was in both periods an area of subsidence, where sediments of a eugeosynclinal character were deposited. Only the Barrandian basin of Bohemia had a similar development.

Silurian

The Silurian beds in continental Europe are dominated by graptolite shales. At the beginning of the period the sea transgressed over wide areas of low relief, and the amount of terrigenous material in the sediments is low in most areas. Limestones occur in some areas, but sandy or greywacke beds are rare, except in the Harz, parts of the Eastern Alps and the Pannonian basin of Hungary, where diabase tuffs also are found, and where the thickness of the Silurian sediments may reach 2000 m (Balogh and Körössy, 1974).

Perhaps some of the undated metamorphic rocks in central and southeastern Europe may provide information giving a more detailed picture of both the Silurian and the other periods in this part of Europe.

In Great Britain sedimentation continued in the three geosynclinal areas of the Ordovician, and a few thousand metres of mostly arenaceous and argillaceous sediments accumulated. Volcanic rocks are almost entirely absent. In the Late Silurian the sea became shallower, and as a rule there is a gradual transition to the continental beds of the Downtonian and Devonian.

The only complete Silurian section in Scandinavia is found in the Oslo area, where the alternation of shales and limestones continued until the sedimentary sequence changed from marine to continental during the Ludlovian. As during previous orogenic phases, the directions of transport and the distribution of sedimentary facies indicate emerging land to the north and west. This time, however, a major land mass was formed.

Outside the Oslo area fossils from the Llandovery are found in the eastern

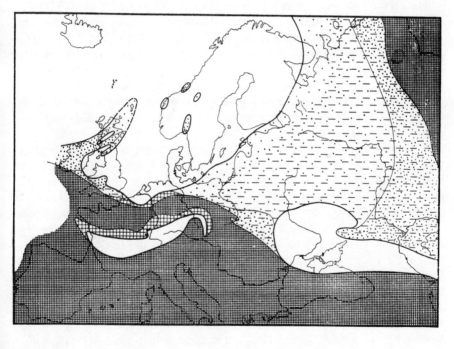

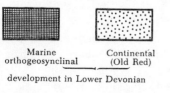

Marine orthogeosynclinal	Continental (Old Red)	Marine orthogeosynclinal	Continental (Old Red)	Marine intercalations in Old Red
development in Lower Devonian		overlapping development in the higher Devonian		

Fig. 14. Palaeogeography of the Devonian in Europe. (After Brinkmann, 1969, Fig. 12)

eugeosynclinal sequence both in the Trondheim area and in western Norway. In the nappes of northern Sweden, Wenlockian fossils have been found in the miogeosynclinal sequence.

If we compare sedimentation during the Silurian in the two branches of the European Caledonides, the differences are less conspicuous than in the Cambrian and Ordovician. Towards the end of the Silurian, however, their modes of development parted completely (Fig. 14). The branch through central Europe continued as a marine geosynclinal area, which evolved into the Variscan geosyncline, while the northeast trending geosynclinal area emerged as a continent where several basins were filled with continental deposits during the following period.

Orogenic Activity

I have tried to outline the main features of development in the European geosynclinal area, especially regarding sedimentation and volcanic activity. Let us now consider the orogenic activity in that area during the Cambro–Silurian periods, as it is manifested in folding, overthrusting, metamorphism and intrusion.

The idea of orogenic phases was especially advocated by Stille (1924) who thought that orogenic movements occurred in relatively short periods, that they affected relatively narrow belts of the Earth's crust, and that they were contemporaneous over wide areas, perhaps throughout the world. Although his work greatly stimulated the study of orogenies, it is now widely admitted that his hypotheses must be modified, especially regarding the synchronous occurrence of orogenic phases. Neither palaeontologic nor radiometric dating is sufficiently accurate to permit a detailed correlation in time of sedimentary series which were deposited in separate basins.

It may also be difficult to determine whether a phase is orogenic or epeirogenic. Most of the orogenic phases which have been postulated in Scandinavia are based upon breaks in sedimentation, in particular the occurrence of conglomerates. If no angular unconformity or no change in metamorphic grade at the contact can be proved, one may sometimes wonder whether the break in sedimentation might be caused by epeirogenic movements. The question is particularly pertinent if the break in sedimentation is recorded in the foreland area only.

I have already mentioned the problem of when the Caledonian orogeny began. In the Metamorphic Map of Europe the boundary between Caledonian and pre-Caledonian metamorphism is placed at 500 Ma or at the beginning of the Ordovician. Being a Scandinavian geologist, I am inclined to include all movements which have affected Eocambrian and Cambro–Silurian rocks before the Late Devonian, and I shall do so here, without discussing when the Caledonian orogeny really began.

Stille (1948) used the name Assyntian for a world-wide Algonkian orogeny. A late stage of this orogeny has been called Cadomian in western Europe and Baikalian in eastern Europe. Some geologists consider the Baikalian to be a later phase of the Cadomian (Sandulescu *et al.*, 1974).

According to the Metamorphic Map of Europe (Fig. 15) radiometric ages of

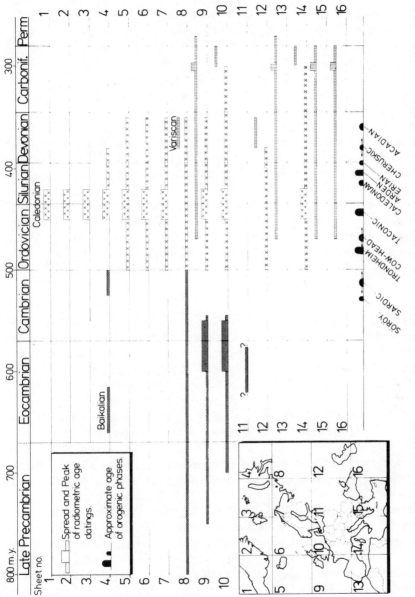

Fig. 15. Late Precambrian and Palaeozoic orogenies in Europe. Based on Metamorphic Map of Europe (1973), except for position of orogenic phases

metamorphism assigned to this orogeny vary, in western Europe from 750–545 Ma, in central Europe from 700–545 Ma and in the Urals from 800–500 Ma. In western and central Europe the peak of matamorphism lies between 550–600 Ma, or in the latest Precambrian and Early Cambrian times. These figures mean that the Cadomian or Baikalian orogeny lasted for about 200 Ma, or about as long as the Caledonian orogeny from the Middle Cambrian to the Late Devonian. During this long period there were several phases of activity. Thus in the Celtic Cycle area there is evidence of folding and/or metamorphism at 680, 640, 610 and 590 Ma, or four times in 90 Ma (Bennison and Wright, 1969).

It would be interesting to try to compare what is known about the activity during this orogeny with what is known about the Caledonian orogeny, but I shall not attempt to do so. I would like to emphasize, however, that while folding and metamorphism were active in a broad belt from Brittany to the Black Sea, there is little evidence of orogenic activity north of the Celtic Cycle area in what later was to become the northwestern branch of the Caledonian geosynclinal area. In southern Norway repeated faulting on old NNW-trending lines of weakness produced the basins where the Eocambrian sediments accumulated. The first two periods of faulting occurred before the deposition of the tillite, which, if it is contemporaneous with the tillite in Finnmark, is probably about 670 Ma (Pringle, 1972). Fig. 16 shows the areas of Caledonian folding in Europe.

Continental Europe

If we consider the Caledonian geosynclinal area in Europe as a whole there was no long and quiet period between the Cadomian or Baikalian and the Caledonian orogeny. It may, in fact, be difficult to determine the boundary between them. According to the Metamorphic Map of Europe the metamorphic ages indicate an interval of 45 Ma covering the Middle and Late Cambrian. In Fig. 17 I have tried to summarize the available information of orogenic activity in the area south of the East European platform. The time scale is largely based upon the Geologic Time Table which was compiled by van Eysinga (1972) but some details have been tentatively added.

In continental Europe evidence of intrusion and/or metamorphism during the Cambrian has been found in the East and South Carpathians of Romania (Sandulescu and Bercia, 1974; Sandulescu et al., 1974), Yugoslavia (Karamata, 1974) and the southern part of East Germany (Hirschmann, 1966; Watznauer, 1965).

Near the boundary between the Cambrian and the Ordovician, a phase which has been considered to be orogenic is recorded in several areas of continental Europe. Various names for this phase have been used, but Sardic is probably most common. It seems to be recorded mostly as a break in sedimentation, and it might perhaps be a matter for discussion whether it should be called an orogenic phase.

The orogenic activity in continental Europe during the Ordovician and Silurian has been a subject of discussion for many years. Stille (1924) and others found evidence of various orogenic phases, which were correlated with those in Great Britain and Scandinavia. Research during the last few decades has raised increas-

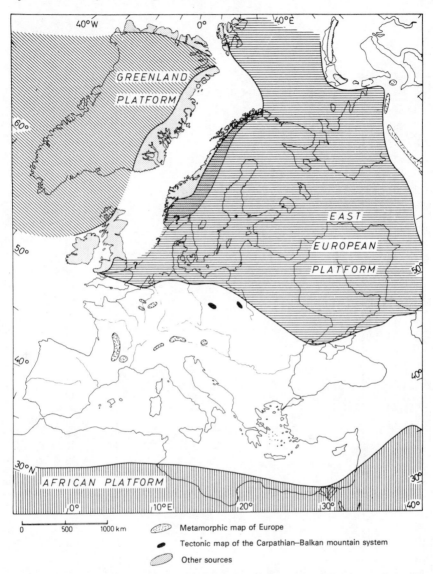

Fig. 16. Areas of Caledonian folding in Europe

ing doubts about the orogenic character of these phases. Brinkmann (1969), Franke (1968) and Lotze (1971) emphasized the epeirogenic character of the movements in central Europe. Radiometric age determinations in recent years indicate, however, the existence of Late Ordovician granite and metamorphism in the Serbo–Macedonian mountains (Karamata, 1974), and of Late Ordovician and Late Silurian metamorphism in the Tokaj mountains in Hungary (Balogh and Körössy, 1974). In Poland three phases of folding are reported from the Gory Swietokrzyskye mountains (Znosko, 1974), two in the Early Ordovician and one during the Ardennian phase.

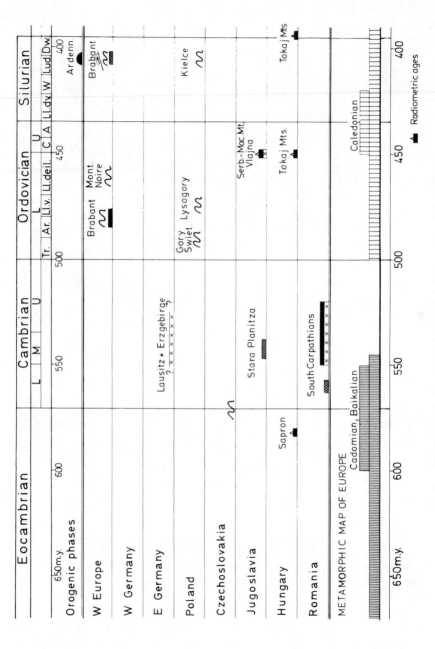

Fig. 17. Events in continental branch of European Caledonides

In western Europe the Brabant Massif provides evidence of Early Ordovician folding, metamorphism, overthrusting towards the north and intrusion of gabbro during the Ardennian phase (Lotze, 1971). In the Montagne Noire in France, folding is recorded in the Middle Ordovician (Gèze, 1960).

Additional evidence of the Caledonian orogenic activity in continental Europe may be found in the Metamorphic Map of Europe. Fig. 18 shows the areas which

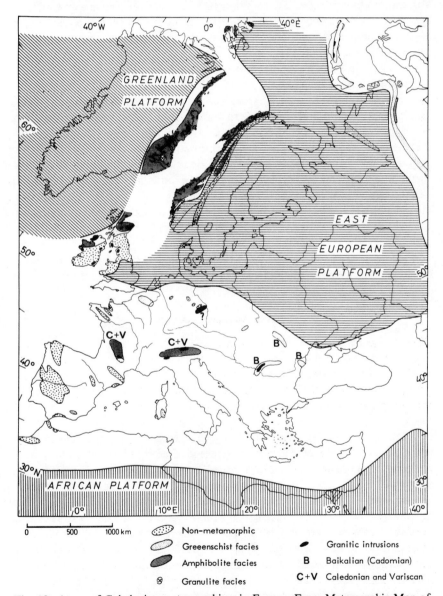

Fig. 18. Areas of Caledonian metamorphism in Europe. From Metamorphic Map of Europe (1973) and other sources

have been subjected to Caledonian metamorphism, as well as areas of Baikalian metamorphism in southeastern Europe. Metamorphism in the greenschist facies is found in the Ardennes–Brabant area, in the south of France and in Hungary, while Baikalian metamorphism in the greenschist facies is found in three areas in Romania. In the south of France and in the central and eastern Alps, areas in amphibolite facies have for the greater part been through both Caledonian and Variscan metamorphism, but some parts of these areas are mapped as having been through the Caledonian orogeny only. Sassi *et al.* (1975) reported evidence of 'a very important and complex event of Ordovician age, whose traces are detectable in the whole Alps.'

If we look at the spread in radiometric ages from the Variscan orogeny (Fig. 15) we find that on maps no. 9, 13, 15 and 17 the Variscan metamorphism may have begun at 470 Ma or during the Middle Ordovician. These figures indicate that in southwestern and southeastern Europe part of the metamorphism which has been called Variscan may in fact be Caledonian.

If we try to summarize the available information about the Caledonian orogeny in continental Europe (Fig. 18) we find areas of unmetamorphic rocks in the north, west and south. They probably border on a zone of greenschist facies metamorphism which in its central parts has areas in amphibolite facies. Migmatization may have occurred in some areas; granites cover minor areas and gabbros are comparatively rare; ultrabasic intrusives are almost lacking. If folding occurs it is moderate, and overthrusting is rare. The horizontal compression seems to have been moderate, if the whole area is considered. In the Serpont massif in the Ardennes Richter (1961) calculated a shortening of 30 per cent.

Northwestern Europe

When we turn to the northwestern branch of the Caledonian geosynclinal area we find that the non-metamorphic Cambro–Silurian in western Europe continues northward through Great Britain to the Highland Boundary fault. The folding is stronger than in central Europe. Dewey (1969) has calculated the shortening to be between 50 and 80 per cent. Several granites have intruded the non-metamorphic rocks.

Northwest of the Highland Boundary fault most of the rocks are in amphibolite facies and migmatization is widespread. Deformation has been extremely complicated. Detailed studies have revealed up to eight stages of deformation (Bennison and Wright, 1969) and to correlate them with the known periods of deformation is no easy task. Altogether nine orogenic phases have been postulated (Fig. 19), but in recent years there has been a tendency to reduce this number. Bennison and Wright state that the movements may probably 'be resolved into at least two main periods of polyphase folding: Early Ordovician metamorphism and folding in the Highlands of Scotland and the northern part of Ireland, and post-Silurian folding in the southern part of Great Britain'.

The Moine thrust has been world famous for more than 80 years, but estimates of the amount of displacement are difficult to find in the literature. Perhaps a few

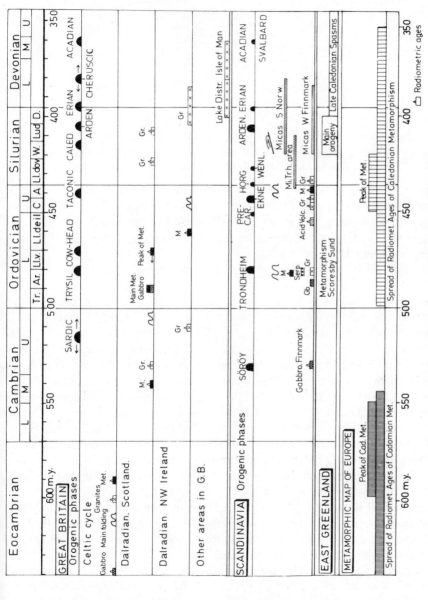

Fig. 19. Events in NW-branch of European Caledonides

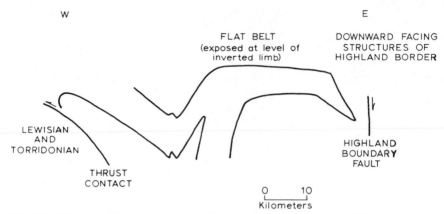

Fig. 20. Diagrammatic cross-section to illustrate regional structure of Iltay nappe complex in Dalradian of southwestern Scotland (after Roberts, 1967)

tens of kilometres would be a reasonable figure. The movements were towards the northwest but a section by Roberts (1967) across the Highlands (Fig. 20) shows that the southeastern part of the Dalradian verges towards the Highland Boundary fault. Although the compression has been considerable there seems to have been no large-scale overthrusting.

In East Greenland a belt of greenschist facies metamorphism appears below the inland ice near the border of the platform. In the coastal areas amphibolite facies prevail. Migmatization is widespread, and Haller (1970) has described the position and movements of the migmatite front. He considered the main Caledonian orogeny to have occurred between 420–400 Ma ago or in the Late Silurian, perhaps corresponding to the Ardennian phase. According to Henriksen (in press) the metamorphism in the Scoresby Sund area, where it reached granulite facies conditions, is, however, of Early Ordovician age. The belt of greenschist facies rocks has been thrust to the west, probably for a moderate distance. Several plutons of granite occur.

In Spitzbergen the Caledonian metamorphism was partly in greenschist facies and partly in amphibolite facies. Some bodies of granite were intruded. There was folding and faulting, but no overthrusting has been recorded.

The most conspicuous feature of the Scandinavian Caledonides is the nappe structure. In Sweden it was discovered by Törnebohm in the 1880s; in Norway it was generally recognized 50 years later. It is now clear that outside the foreland area the vast majority of rocks are allochthonous or parautochthonous. This fact, and the scarcity of fossils, are the main reasons why so many problems are unsolved or proposed solutions are disputed.

A generalized map of the metamorphic zones (Fig. 18) shows a non-metamorphic foreland area, a belt of greenschist facies metamorphism, and in the coastal areas a belt of amphibolite facies. Granulite facies conditions have been approached in some areas but not reached.

An interesting feature of the Scandinavian Caledonides is the zone of Precambrian rocks in or near the coastal areas from 59 to 70°N. (Fig. 21). Most of

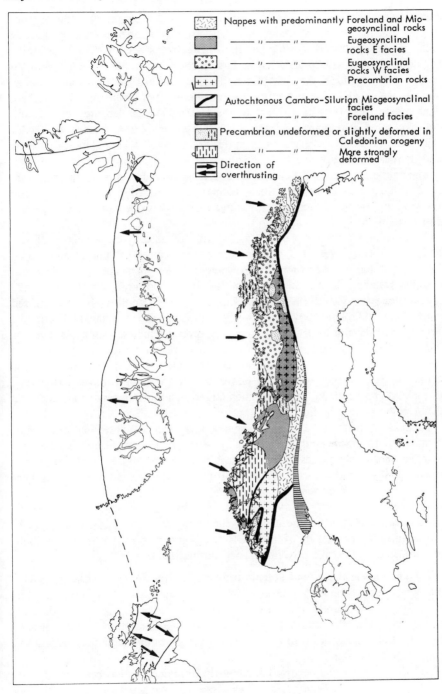

Fig. 21. Nappes in the Scandinavian Caledonides

them have recrystallized in Caledonian times under amphibolite facies conditions, and the structural trends generally strike NNE, while comparable rock units in the undisturbed Precambrian generally strike NNW. Caledonian migmatization is widespread, but it has rarely affected the Cambro–Silurian rocks. Radiometric K/Ar determinations give evidence of Caledonian deformation, while whole rock Rb/Sr determinations have given ages in southern Norway up to 1800 Ma. In part of the Lofoten–Vesterålen area in northern Norway the Caledonian influence is hardly detectable. According to Taylor (1975) the Pb/Pb age of a gneiss from this area is 3460 Ma, making it the oldest known rock in Europe.

The nappe area lies to the east of the coastal gneiss ridge. A large number of individual nappes have been described, as might be expected in an area about 1700 km long and up to 500 km wide. Various authors have grouped them together, into three to five units.

There is strong evidence for an easterly to southeasterly movement of all tectonic units. The lowest and easternmost nappes consist of unmetamorphosed rocks which have moved only short distances. The higher nappes contain metamorphic Cambro–Silurian and Precambrian rocks and have moved greater distances. Estimates vary considerably. Nappes in central southern Norway moved more than 100 km above undisturbed Precambrian basement. From the Trondheim area northward, movements of several hundred kilometres, even as much as 1000 km, have been postulated but there is little reliable evidence for such estimates. It is, however, quite clear that the eastward movements of the nappes in Scandinavia were much larger than the westward movement in East Greenland and Scotland, and also much larger than the nappe movements during the Alpine orogeny in Europe.

As in Great Britain nine orogenic phases have been postulated but only five of them seem to correspond in time to those in Scandinavia, bringing the total number of orogenic phases or disturbances on both sides of the North Sea to thirteen (Fig. 19). Some of them may be epeirogenic, but it seems that in five of them orogenic activity is evident.

During the last few years various attempts have been made at constructing plate tectonic models for the development of the Scandinavian Caledonides. In this connection it might be of interest to try to summarize what is known about the orogenic phases and what happened during them.

(1) The earliest phase, which I tentatively call here the *Söröy* phase, is marked by the intrusion of the Hasvik gabbro in Söröy, Finnmark 530 Ma ago (Pringle and Sturt, 1969). It is doubtful if other evidence of this phase is known in Scandinavia, and it lies within the long and quiet period of the alum shale sea in southern Norway, but it coincides with the intrusion of granites and metamorphism in the Dalradian of Scotland.

(2) The second event occurred in the Early Ordovician. It has been called either the *Trondheim* or the *Trysil* phase. The intense volcanic activity in the Arenig was followed by intrusion of serpentinites, gabbros, quartz diorites and granites during a period of metamorphism and folding.

(3) The Middle Ordovician seems to have been quiet but there were periods of unrest in both the Late Ordovician and the Early Silurian. Conglomerates were formed which show a similar lithology in western Norway, the Trondheim area and northern Sweden, but whether the conglomerates are contemporaneous is not known. The conglomerate following the Late Ordovician *Ekne* disturbance contains pebbles of Cambro–Ordovician rocks in a matrix of greywacke, while the conglomerate following the Early Silurian *Horg* disturbance has pebbles predominantly of quartz or quartzite in an arenaceous matrix. Several age determinations give evidence of intrusion of granites and of metamorphism in Late Ordovician and Early Silurian times (Fig. 19).

(4) The Late Silurian *Ardennian* disturbance has been considered previously to be the main Caledonian orogenic phase in Scandinavia (for example Vogt, 1929). It has been supposed to include downfolding of Cambro–Silurian supracrustal rocks to great depths, deformation, metamorphism, migmatization, granitization and overthrusting. Evidence obtained in recent years indicates that this opinion must be revised.

The Ardennian phase seems to have been mainly a period of overthrusting. In Sweden the youngest fossiliferous rocks in the nappes are from the Early Wenlock; in Norway they are from the Middle or Late Llandovery. As the fossils in Sweden are found in all nappes except the uppermost ones, which contain no fossils, it seems likely that the thrusting of all nappes in that area was post-Lower Wenlockian. Various authors have assumed that overthrusting also occurred during earlier phases, even in the Ordovician, but the evidence is not conclusive. No estimate has, to my knowledge, been made of the amount of shortening resulting from either the overthrusting or the earlier phases of folding.

(5) The last of the Caledonian orogenic phases recorded in Scandinavia is the *Acadian* or *Svalbardian* phase, which involved gentle folding, faulting and some overthrusting, but no metamorphism of the Middle Devonian sediments.

If we try to generalize the available information about Caledonian orogenic activity in Great Britain, East Greenland and Scandinavia, the following pattern seems to emerge.

(1) A *Late Cambrian* phase of folding, intrusion and metamorphism in the Scottish Highlands, northern Ireland and West Finnmark.

(2) An *Early Ordovician* phase of folding, intrusion and metamorphism in the Scottish Highlands, northern Ireland, East Greenland and Scandinavia, especially in the coastal areas of western Norway.

(3) A *Late Ordovician to Early Silurian* period of intrusion and metamorphism, including migmatization, especially in Scandinavia and East Greenland.

(4) A *Late Silurian* phase of folding and overthrusting.

This generalization is tentative and it will certainly need revision. It is offered as a topic for discussion, because it would be useful for our future work on the European Caledonides if we could reach some kind of agreement on what really happened in those times.

Concluding Remarks

The Devonian folding was moderate to weak in Great Britain as well as in East Greenland and Norway. In post-Devonian times there is no evidence of compression. North–south trending Carboniferous faults in East Greenland give evidence of some tension in the Caledonian direction of compression. During the Mesozoic, conditions of sedimentation were similar east of Greenland and in the North Sea. Major horizontal movements did not occur until the beginning of the Tertiary, when the separation of Greenland and Scandinavia began.

In continental Europe the post-Caledonian development was entirely different. In the same general area where the Cadomian or Baikalian and the Caledonian orogenies had been active, the end of the Caledonian era marked the beginning of the Variscan era, which in turn was succeeded by the Alpine era.

What is the cause of this fundamental difference in development between the two branches of the European Caledonides? In pre-plate tectonic days the idea of an area becoming 'totgefaltet' had some support. If an area was strongly folded and metamorphosed, the crust would become so solid that it could not again become part of an orogen, but was added to one of the stable blocks. It was, however, difficult to understand why the Precambrian in the coastal areas of Norway should be less stable than the corresponding rocks further east. And why were three orogenies in central Europe, the Cadomian–Baikalian, the Caledonian and the Hercynian not capable of producing a stable crust in the area which later was subjected to the Alpine orogeny (Fig. 22)?

In recent years one has sometimes felt that the theory of plate tectonics is supposed to explain all of our major tectonic problems. This theory has without doubt greatly improved our understanding of the processes both in and below the Earth's crust. But 200 years of geological research should have taught us that nature is more complicated than we can imagine in our hypotheses and theories. It should also have taught us that nature can arrive at the same or very similar results in more than one way. The application of plate tectonic theory to the Alps has been criticized by Trümpy (1971), who found that 'in the Alps, the stratigraphical sequence is better compatible with the hypothesis of oceanization (replacement of the lower part of the crust by denser matter) than with the mechanism of ocean-floor spreading according to the Atlantic model'.

We know much less about the Caledonian than about the Alpine orogeny in Europe. Some attempts have been made to apply the theory of plate tectonics to the European Caledonides with, as one might expect, different results. I shall not comment upon those attempts now. In my opinion the need today is for an appraisal of where we stand, what we know about the Caledonides, and in what directions our research should be aimed in order to fill the most important gaps in our knowledge. If we could do so we might considerably improve the fundamental knowledge upon which our attempts to understand the history and development of the European Caledonides must be based.

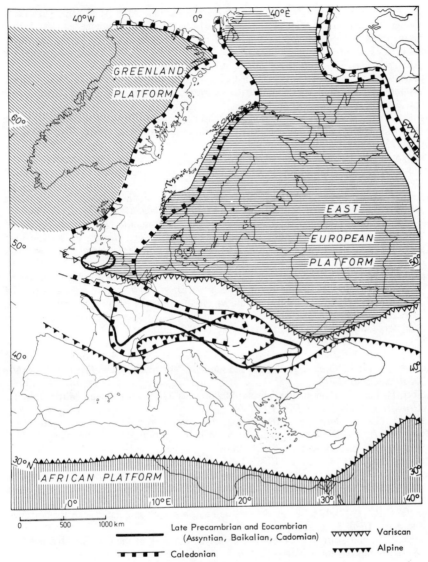

Fig. 22. Orogenic belts in Europe

References

Balogh, K. and Körössy, L. (1974). 'Hungarian Mid-Mountains and adjacent areas', *Tectonics of the Carpathian Balkan Regions*, Geological Institute of Dionýs Stúr, Bratislava 391–407

Bennison, G. M. and Wright, A. E. (1969). *The Geological History of the British Isles*, Edward Arnold, London

Björlykke, K. (1974). 'Depositional history and geochemical composition of Lower Palaeozoic epicontinental sediments from the Oslo region', *Norges geol. Unders.*, 305

Bogdanoff, A. A., Mouratov, M. V. and Schatsky, N. S. (eds.) (1964). *Tectonics of*

114   Anders Kvale

Europe. Explanatory note to the International Tectonic Map of Europe. (Int. Geol. Congr, Comm. Geol. Map of the World), 1–260 (Russian language edn.)

Brinkmann, R. (1969). *Geologic evolution of Europe*, Ferdinand Enke Verlag Stuttgart.

Bullard, E., Everett, J. A. and Smith, A. G. (1965). The fit of the continents around the Atlantic', *Phil. Trans. R. Soc. Lond.*, A258, 41–75

Dalziel, I. W. D. (1969). 'Pre-Permian history of the British Isles—a summary, North Atlantic geology and continental drift—a symposium. *Am. Ass. Petrol. Geol. Mem.*, 12, 5–31, Tulsa, Oklahoma

Dewey, J. F. (1969). 'Structure and sequence in paratectonic British Caledonides', North Atlantic geology and continental drift–a symposium. *Am Ass. Petrol. Geol. Mem.*,12, 309–35. Tulsa, Oklahoma

Eysinga, F. W. B. van (1972). *Geological Time Table*, Elsevier, Amsterdam

Franke, D. (1968). 'Zum Problem der kaledonischen Gebirgsbildung', *Grundriss d. Geol. d. DDR*, 1, 153–6, Akademie-Verlag, Berlin

Gaertner, H. R. von (1960). 'Uber die Verbindung der Bruchstücke des kaledonischen Gebirges im nördlichen Mitteleuropa', *XXI Intern. Geol. Congr., Pt. XIX*, 96–101

Gèze, B. (1960). 'L'orogénèse calédonienne dans la Montagne Noire (sud du Massif Central Français) et les regions voisines', *XXI Intern. Geol. Congr.*, Pt. XIX, 120–5

Haller, J. (1970). Tectonic map of East Greenland (1:500,000), *Medd. om Grönland*, 171, 5

Henriksen, N. (1975). 'The southern part of the Caledonian fold belt in East Greenland, with special reference to the Scoresby Sund Region', *Bull. geol. Soc. Denmark*, 24

Hirschmann, G. (1966), 'Assyntische und variszische Baueinheiten im Grundgebirge der Oberlausitz (unter spezieller Berücksichtung der Geologie des östlichen Görlitzer Schiefergebirges)' *Freiberger Forsch. -H.*, C 212, Leipzig

Karamata, S. (1974). 'Evolution of Magmatism in Jugoslavia', *Tectonics of the Carpathian Balkan Regions*, Geological Institute of Dionýz Stúr, Bratislava, 354–7

Kent, P. E. (1967). 'Outline geology of the southern North Sea Basin', *Proceed. Yorkshire Geol. Soc.*, 36, 1–22

Kölbel, H. (1968). 'Regionalgeologische Stellung der DDR im Rahmen Mitteleuropas', *Grundriss d. Geol. d. DDR*, 1, 18–66

Lotze, F. (1971). *Dorn-Lotze: Geologie Mitteleuropas*, E. Schweizerbart, Stuttgart

Martinson, A. (1974). 'The Cambrian of Norden', *Cambrian of the British Isles, Norden, and Spitsbergen*, Holland, C. H. (ed.) Wiley, London, 185–283

Metamorphic Map of Europe. (1973). Leiden/UNESCO, Paris

Pozaryski, W., Malkowski, J. and Jankowski, J. (1965). 'Distribution of short-period geomagnetic variations related to tectonics in Central Europe', *Roczn. Polsk. towarz. geol. Ann. Soc. Géol. Pol.*, 35, 1, 97–102

Pringle, I. R. (1972). 'Rb/Sr age determinations on shales associated with the Varanger Ice Age', *Geol. Mag.*, 109, 6

Pringle, I. R. and Sturt, B. A. (1969). 'The age of the peak of the Caledonian orogeny in West Finnmark, north Norway', *Norsk. geol. Tidsskr.*, 49, 435–6

Ramsay, D. M. (1971). 'Stratigraphy of Söröy', *Norges geol. Unders.*, 269, 314–17

Richter, D. (1961). 'Zur Baugeschichte der Ardennen', *Geol. Rundsch.*, 51, 574–600

Roberts, J. F. (1967). 'The formation of similar folds by inhomogeneous plastic strain, with reference to the fourth phase of deformation affecting the Dalradian rocks of the Southern Highlands of Scotland', *Jour. Geol.*, 74, 831–55

Săndulescu, M. and Bercia, I. (1974). 'The East Carpathians. The Crystalline-Mesozoic zone', *Tect. of Carp. Balk. Reg.*, 240–53

Săndulescu, M., Năstăseanu, S. and Kräutner, H. G. (1974). 'The South Carpathians', *Tect. of Carp. Balk. Reg.*, 264–76

Sassi, F. P., Schonlaub, M. P. and Zanferrari, A. (1975). Early history of the Eastern Alps', *Abstracts of Keynote addresses and short communications. Meeting of European geol. Soc.* Reading

Siedlecka, A. and Siedlecki, S. (1972). 'Lithostratigraphical correlation and sedimentology of the Late Precambrian of Varanger Peninsula and neighbouring areas of East Finnmark, northern Norway', *XXIV Intern. Geol. Congr.*, **6**, 346–58

Siedlecka, A. (1975). 'Late Precambrian stratigraphy and structure of the northeastern margin of the Fennoscandian shield (East Finnmark–Timan region', *Norges geol. Unders.*, **316**, 313–48

Sorgenfrei, T. and Buch, A. (1964). 'Deep tests in Denmark 1935–1959', *Geol. Surv. Denmark*, **III R**, No. 31

Stille, H. (1924). *Grundfragen der vergleichenden Tektonik*, Gebr. Borntraeger, Berlin

Stille, H. (1948). 'Die assyntische Ära und der vor-, mit-, und nach-assyntische Magmatismus', *Z. dt. geol. Ges.*, **98**, 152–156

Störmer, L. (1967). 'Some aspects of the Caledonian geosyncline and foreland west of the Baltic Shield', *Q. Jl. geol. Soc. Lond.*, **123**, 183–214

Strand, T. (1972). 'The Norwegian Caledonides', *Scandinavian Caledonides*, Pt. 1, Wiley–Interscience, London

Sturt, B. A. and Ramsay, D. M. (1965). 'The alkaline complex of the Breivikbotn area, Söröy, northern Norway', *Norges geol. Unders.*, 231

Svoboda, J., Havliček, V. and Horný, R. (1966). The Barrandian Basin. Regional geology of Czechoslovakia, *Geol. Surv. Chechoslov.*, Prague, **1**, 281, 341

Taylor, P. N. (1975). 'An early Precambrian age for migmatitic gneisses from Vikan i Bö, Vesterålen, North Norway', *Earth and Planet. Sci. Lett.*, **27**, 35–42

Trümpy, R. (1971). 'Stratigraphy in mountain belts', *Q. Jl. geol. Soc. Lond.*, **126**, 293–318

Vogt, T. (1929). 'Den norske fjellkjedes revolusjonshistorie', *Norsk geol. Tidsskr.*, **10**, 97–115

Watznauer, A. (1965). 'Stratigraphie und Fazies des erzgebirgisches Kristallin im Rahmen des mitteleuropäisches Varistikums', *Geol. Rdsch.*, **54**, 853–60

Znosko, J. (1974). 'Polish Carpathian foreland', *Tectonics of the Carpathian Balkan Regions*, Geological Institute of Dionýz Stúr, Bratislava, 431–42

Meso-Europa

That part of the continent unaffected by major
tectonism since the end of Late Palaeozic times

THE TECTONIC EVOLUTION OF VARISCAN
MESO-EUROPA

W. KREBS
*Institut für Geologie und Paläontologie der Technischen Universität,
D-33 Braunschweig, Pockelsstr. 4, Federal Republic of Germany*

Abstract

Krebs, W. (1976). 'The Tectonic Evolution of Variscan *Meso-Europa*, in Ager, D. V. and
Brooks, M. (edit.), *Europe from Crust to Core*, Wiley, London.

The concept of a homogeneous step-like geotectonic development of the Central European basement, advocated by Stille, has become more and more doubtful in the last
decade. A widespread independent Assyntian orogen never existed in central Europe. The
metamorphic complexes in Europe (Spain, Armorican Massif, Massif Central, Vosges,
Black Forest, Erzgebirge, Moldanubian Massif and others) were already Precambrian
stabilized cores within the Late Precambrian–Palaeozoic geosynclinal evolution. Also a
Caledonian deformed *Meso-Europa* comprising large parts of central Europe never existed. Alpinotype Caledonian deformations are known only in isolated core structures.
   The anticlinal areas and massifs are characterized by a steady, higher heat flow,
documented by metamorphism and anatexis. Local unconformities, incomplete
sedimentation and thinned sedimentary sections indicate erosion and/or non-deposition of
the rising core zones, often festoon-like, surrounded by shallow-water limestones and
coarse-grained clastic sediments. Continuous, thick and only weakly metamorphosed
Palaeozoic sequences, however, are restricted to narrow synclinal areas in front of or
between the anticlinal structures (for example, Armorican Massif, Saxothuringian zone).
A migration of Palaeozoic troughs during geosynclinal evolution towards the foreland occurs only in the external parts. Here, narrow foredeeps with paralic to limnic Late Carboniferous molasse sediments characterize the final phase of geosynclinal subsidence
(Asturian–Cantabrian zone, south Portuguese zone, Subvariscan zone, Moravosilesian
zone). In the central European Palaeozoic basement some larger, more or less elongate,
crystalline belts display polarity axes (intermediate massifs). In the adjoining areas, thrust
planes and axial fold planes diverge outwards from the crystalline zones. There are
different opinions concerning the connection between these polarity axes in Western
Europe. Besides the more elongate, bilaterally symmetrical intermediate massifs, in the
Iberian and German Palaeozoic, isolated diapir-like extrusions of the Precambrian basement are also known. The folds surrounding the domes also show opposite vergences
(Münchberg, Granulitgebirge). Carboniferous post-kinematic granites intrude into the
predestined rises and anticlinal structures. Due to the vertical uplift of the older core zones,
nappes and gliding masses move down from their flanks in the direction of the subsiding
basins.
   Parts of the uplifted crystalline belts are inverted to graben-like, intramontane molasse
basins with limnic, often coal-bearing Permo–Carboniferous sediments (Saar–Nahe trough,
Saale trough). Other subsequent basins follow narrow older synclines (Laval syncline) or
lineaments (Sillon Houiller in the Massif Central, Elbe valley in the Saxothuringian zone).
The post-kinematic stage is characterized by tension, vertical movements, local folding
due to vertical ascent of magmas and widespread rhyolitic–andesitic–basaltic 'subsequent' volcanism.

119

In recent years there have been repeated attempts to interpret the Devonian–Carboniferous palaeogeography of Europe in particular in the content of plate tectonics. Worth mentioning is the hypothesis of a Palaeozoic ocean between the Rhenohercynian and Saxothuringian zone with one or two oppositely dipping subduction zones. Spilite and greenstones in SW England and Germany and Late Devonian–Early Carboniferous deeper water sediments are interpreted as remnants of a former Mid-European ocean basin.

There are no indications of a Palaeozoic deep ocean with 'oceanic' crust in Central Europe. The whole basement of the Palaeozoic in Europe is undoubtedly ensialic. The Devonian and Early Carboniferous effusive and intrusive basalts do not represent either an ophiolite suite or an equivalent of modern oceanic crust. Ultrabasic rocks are very scarce and are restricted to the rims of the squeezed out basement diapirs (Münchberg, Granitgebirge, Eule). The pillow basalts and spilites are concordantly interbedded in marine Palaeozoic sediments and range in age from Late Proterozoic to Early Carboniferous. Their thickness attains more than 300 m in only a few cases. There are no linear 'greenstone belts' or 'ophiolite sutures' from SW England to the Sudeten Mountains, but the basalts fringe the margins of the isolated core zones. On the hinge-line between the rising core zones and the contemporaneously subsiding basins deep-seated faults originated on which basic melts of mantle or deep crustal origin ascended. The Permo–Carboniferous andesitic–rhyolitic volcanism does not form a uniform linear belt like an island arc, but is bound to anticlines, collapsed subsequent basins, older synclines or lineaments.

No reconstructions of former Palaeozoic plate boundaries or subduction zones in western and central Europe correspond with the well-known geological facts in the field. On the contrary, during Palaeozoic geosynclinal evolution in central Europe the crystalline cores and local diapiric structures, with repeated metamorphism and plutonism, are a result of vertically ascending mantle diapirs. Analogous to the marginal sinks on the flanks of salt diapirs, subsiding furrows were formed contemporaneously with the rising of granitic melts into uppermost crustal levels. The driving force of folding, thrusting, allochthonous gliding masses and nappe transport was the potential energy between the rising and expanding core zones and the synchronously subsiding adjacent troughs.

Delimitation and Subdivision

Stille (1924) divided Europe into *Eo-Europa, Palaeo-Europa, Meso-Europa* and *Neo-Europa. Eo-Europa* was formed by the Precambrian of Fennosarmatia, *Palaeo-Europa* resulted from the deformed Caledonian geosyncline of the British Isles and Norway, *Meso-Europa* from the folding of the Variscan or Hercyian geosyncline in central and southwest Europe. *Neo-Europa* represents the scene of the Alpine orogeny. According to Stille these units are supposed to have enlarged the European continent step by step from north to south.

Stille's units indicate only the coherent areas of the youngest orogenic deformation. *Meso-Europa* is therefore defined as that part of Europe which remained unaffected by major tectonism since the late Palaeozoic (Ager, 1975). The folded, faulted and partly metamorphosed Precambrian and Palaeozoic basement of *Meso-Europa* is covered by thick epicontinental late Palaeozoic to Cenozoic sedimentary series. The post-Variscan cover of *Meso-Europa* lies outside the scope of this paper.

If in the *Meso-European* fold belt the Variscan orogenic deformation finished homogeneously in late Palaeozoic times, there is no uniform basement of the same age. A widespread Assyntian or Cadomian orogen never existed in *Meso-Europa*, nor a Caledonian deformed central Europe comprising large parts of *Meso-Europa*. Alpinotype Caledonian deformations are known only in isolated core structures, for example the Brabant Massif, Ardennes, parts of the Armorican Massif or along the southwestern border of the East European Precambrian platform in Rügen and Pomerania.

The boundary between the Old Red Sandstone-type Devonian deposits and the marine Devonian sequence passes from South Wales to the Brabant Massif. A still open question is the continuation of the boundary between *Palaeo-Europa* and *Meso-Europa* in the subsurface of the southern North Sea and northern Germany. In many palaeogeographic reconstructions, the East European Precambrian platform is even interpreted as being extended as far as the English Midlands (Bailey, 1928). Other maps of the area in question show the existence of Precambrian massifs (North German High, East Elbe Massif).

The spreading of the Devonian Old Red Sandstone facies represents a key for the reconstruction of the Caledonian mountain ranges of *Palaeo-Europa*. Areas with evidence of the Caledonian orogeny (Ireland, Britain, the Brabant Massif, Ardennes, Rügen, Pomerania, Holy Cross Mountains) are always characterized by a transgressive Old Red Sandstone facies in the Devonian. The influence of the nearby Caledonian folded areas is also shown in the interfingering of the continental Old Red Sandstone facies with marine shelf sediments in the Devonian of southwest England and Germany (House, 1975).

Early Carboniferous shelf limestones also seem to typify the Caledonian stabilized areas in northwest and north Europe. According to many authors there is a continuous belt of Early Carboniferous shallow-water limestones from south Ireland through the subsurface of northwestern Germany and southwestern Baltic Sea to Poland (Ziegler, 1975; Schmidt and Franke, 1975). But it is also possible that the Early Carboniferous limestones are restricted to local elevated core zones only (Krebs, 1975a). The southern border of *Meso-Europa* is equivalent to the Alpine front running from the Iberian Peninsula to the Black Sea.

In accordance with the classical work of Kossmat (1927) the central European Variscan fold belt can be divided from north to south into the following zones (Fig. 1):

Subvariscan zone
Rhenohercynian zone
Central German crystalline rise
Saxothuringian zone (including the West Sudetes)
Moldanubian zone
Moravosilesian zone (East Sudetes).

Lotze (1945, 1963) subdivided the Iberian Variscides from north to south into six zones of approximate northwest–southeast trend (Fig. 1):

Cantabrian zone
West Asturian–Leonesian zone

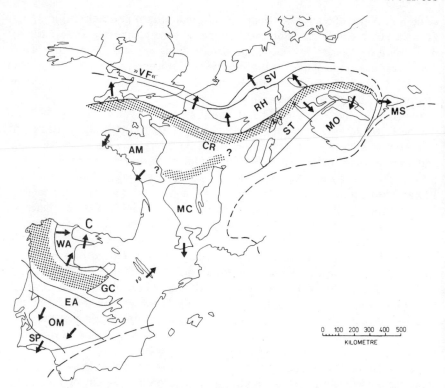

Fig. 1. Variscan zones, crystalline median massifs (stippled) and direction of vergences (arrows) in *Meso-Europa*. 'VF': Variscan Front; SV: subvariscan zone; RH: Rhenohercynian zone; CR: Central German crystalline rise; ST: Saxothuringian zone; MO: Moldanubian zone; MS: Moravosilesian zone; AM: Armorican Massif; MC: Massif Central; C: Cantabrian zone; WA: West Asturian–Leonesian zone; GC: Galician–Castilian zone; EA: East Lusitanian–Alcudian zone; OM: Ossa–Morena zone; SP: South Portuguese zone

Galician–Castilian zone
East Lusitanian–Alcudian zone
Ossa–Morena zone
South Portuguese zone

This subdivision, showing the Galician–Castilian zone as a central metamorphic zone with a gross fan-like outwards vergence pattern, was slightly modified by the work of Matte (1968) and Bard *et al.* (1973).

Main Characteristics of *Meso-Europa*

From the Proterozoic to the Early Carboniferous, the large anticlinal structures in the Variscan fold belt are identical with areas of repeated metamorphism, anatexis, granite intrusion, uplift, erosion and incomplete sedimentation. The large synclines, on the other hand, resulted from mobile subsiding basins with more or less

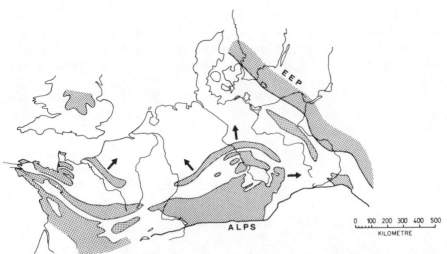

Fig. 2. Precambrian stabilized core zones (stippled) in western and central Europe (after Hoth and Hirschmann, 1970). The stationary troughs are developed between the older core zones. The direction of migration of subsidence and sedimentation in the mobile troughs is indicated by arrows. EEP: East European Platform

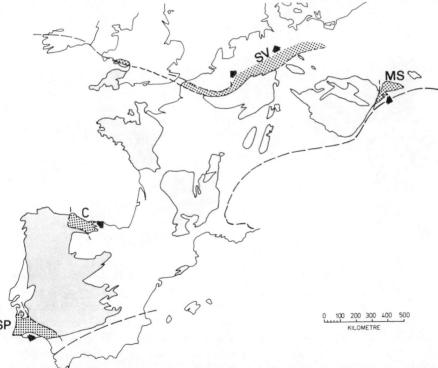

Fig. 3. The mobile flysch-molasse foredeeps in the external parts of the *Meso-European* Variscan fold belt. The arrows indicate the direction of migration. SP: South Portuguese zone; C: Cantabrian zone; SV: Subvariscan zone; MS: Moravosilesian zone

thick and uninterrupted sedimentary series. These sedimentary series are either only weakly metamorphosed or not at all. In *Meso-Europa* the zones of maximum Palaeozoic subsidence are restricted to stationary oblong to oval troughs between the already stabilized Precambrian core zones (Fig. 2) and to mobile external fore-deeps which migrated gradually towards the forelands (Fig. 3).

According to Zwart (1967), regional metamorphism, migmatites and intrusive granites are extremely widespread and abundant throughout the whole Variscan fold belt. The area involved is approximately 1500 × 2000 km in size. Low

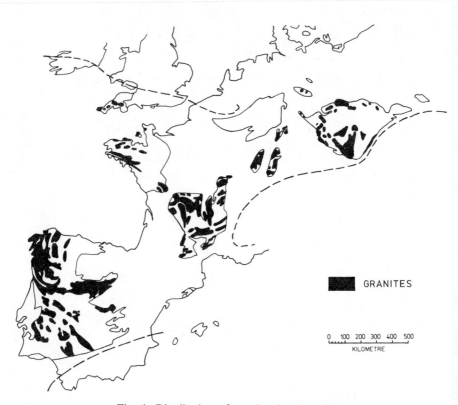

Fig. 4. Distribution of granites in *Meso-Europa*

pressure–high temperature metamorphism is characteristic and many regional metamorphic rocks resulted from the reactivation of older basement rocks.

The widespread granitic intrusions (Fig. 4) are connected with metamorphism and migmatism of varying intensity, especially in the above-mentioned anticlinal areas (for example Massif Central, Vosges, Black Forest, Erzgebirge, Modanu-bian Massif). A culmination of intrusions took place in the Carboniferous, but older granites are also known by radiometric determinations, especially in the latest Precambrian, Ordovician and early Devonian. Most of the Late Carbon-iferous post-orogenic granitic plutons are restricted to older anticlinal structures.

Vergences

According to Stille (1951) and Dvořák and Paproth (1969), the Moldanubian zone belongs to the 'Alemannischer Scheitel' with fan-like vergences directed outwards to the Saxothuringian in the northwest and the Moravosilesian in the southeast. Brause (1970) and other authors, however, have shown that the northern belt of the Saxothuringian zone—*the Central German crystalline rise* (Mitteldeutsche Kristallinschwelle)—forms the main vergence fan in central Europe (Fig. 1). The Rhenohercynian and Subvariscan zones are characterized by northwestward thrusting and folding. Besides local anticlines and some structures, the predominant vergence in the Saxothuringian zone is directed towards the southeast.

The central German crystalline rise is characterized by low-pressure metamorphic rocks, Variscan granitoid intrusives, positive magnetic anomalies and local gravity highs. Large parts of the crystalline rise are inverted into graben-like intramontane basins which are filled with limnic Permo–Carboniferous molasse sediments up to a thickness of 5000 m. The post-orogenic tension in the intradeeps is connected with eruptions and intrusions of acid, intermediate and basic volcanics.

There are two different opinions regarding the western continuation of the central German crystalline rise. Some authors propose a continuation under the Paris Basin to the area south of southwest England (v. Gaertner, 1969; Schoenenberg, 1971). Others, however, believe that this crystalline belt stretches north of the Massif Central into the Armorican Massif (v. Bubnoff, 1930; Aubouin, 1965; Hoth and Hirschmann, 1970).

Another axial crystalline zone, bordered by belts with opposite vergences directed outwards forms the *Galician–Castilian zone* in Spain and north Portugal (Fig. 1). In the Galician–Castilian zone, Precambrian metamorphic rocks and Variscan granites are widespread. The Precambrian basement is partly remobilized or squeezed out to mushroom-shaped circular structures.

Outside the above-mentioned crystalline axes, local anticlines or dome structures show a fan-like arrangement of axial fold planes in the Saxothuringian, Rhenohercynian and Moravosilesian zone (for example, the Sternberk–Horni Benešov structure; Schoenenberg, 1973). The small strip of southward vergence immediately northwest of the Central German crystalline rise in the Rhenish Schiefergebirge (for example southeast of Hunsrück and the Taunus Mountains) is a result of post-orogenic rotation due to the strong subsidence of the adjacent Late Carboniferous intramolasse basin. Locally, the folding of the nappes gives rise to structures of apparently reversed vergence (Julivert, 1971). According to the subduction model, the dip of the Benioff zone is identical with the dip of the thrusts. Consequently any plate tectonic interpretation of *Meso-Europa* should consider the opposite vergences in the different parts of the Variscan fold belt.

Anticlines and Metamorphic Massifs

The large anticlines and metamorphic massifs in the Variscan fold belt have Precambrian cores (Fig. 2). According to Hoth and Hirschmann (1970), these

Precambrian core zones do not represent fragments of older orogenies but formed early stabilized areas during the Proterozoic–Palaeozoic evolution of *Meso-Europe*.

The large anticlines and metamorphic massifs are characterized by a repeated, higher heat flow documented by polyphase metamorphism, anatexis and granitization. Deep erosion levels of the crystalline basement show the typical 'mantled gneiss dome' structure, for example in the Moldanubian massif, eastern Erzgebirge, Böllstein Odenwald. The decreasing heat flow of the steeply dipping thermal gradients caused in the sedimentary cover the formation of concentric metamorphic aureoles that may be traced continuously around local anticlinal structures. In the Thuringian Palaeozoic, 'metamorphic islands' are identical with intra-geosynclinal rises, condensation, sedimentary gaps, local unconformities and pre-orogenic intrusives. These epizonal metamorphic aureoles cut discordantly different stratigraphical levels up to the Dinantian (Hempel, 1974; Meinel, 1974). Finally, the deeper-seated thermal domes are reflected in the thick sedimentary overburden by areas of higher coalification of organic material, higher illite crystallinity, hydrothermal ores and geophysical anomalies. These connections are evident in the Middle Devonian core of the East Sauerland anticline in the northeastern Rhenish Schiefergebirge (Weber, 1972; Wolf, 1972; Paproth and Wolf, 1973) or in the Carboniferous Krefeld dome (Krebs and Wachendorf, 1974).

As already mentioned, the anticlinal areas are marked by uplift, erosion, local unconformities and sedimentary gaps. The central zone of the eastern Erzgebirge was already folded and metamorphosed at the end of the Proterozoic. Cambrian conglomerates, clastics, carbonates and diabases 'surround' the stabilized core zone and were themselves metamorphosed in a later phase (Bernstein *et al.*, 1973). In early Late Devonian time the exposed core of the Hirschberg–Gefell anticline in Thuringia, which shows eroded pre-orogenic granitic intrusions, was festoon-like, surrounded by conglomerates, graywackes and spilitic volcanics (Gräbe *et al.*, 1968).

Post-orogenic Carboniferous and Permian granites were intruded mostly into older anticlinal areas, for example in the Armorican Massif and the Saxothuringian zone. Some of the older anticlinal structures show a post-orogenic inversion to graben-like molasse basins with limnic red-bed sedimentation and Permo–Carboniferous volcanism.

Particularly worth mentioning are elliptical to circular diapir-like dome structures in the Saxothuringian and the Iberian Variscan basement. The Münchberg gneiss massif in northeastern Bavaria and the Granulitgebirge in Saxony are well known. Other examples form the Precambrian Sowie Góry (Eule) massif in Silesia (Morawski, 1973) and more or less circular Precambrian complexes within the Galician–Castilian zone of northwest Spain and north Portugal (Matte, 1968; Ries and Shackleton, 1971).

These mostly mushroom-shaped dome structures consist of medium- to high-grade metamorphic rocks which have been intensively folded, refolded and mylonitised. The widespread granulite and ecologite facies indicate a protrusion of squeezed-up, deep crustal rocks which were later altered by retrograde Variscan metamorphism and mylonitisation.

The metamorphic nuclei are mostly festoon-like, surrounded by ultrabasic to basic rocks. These basic rocks were injected during and after the tectonic emplacement on the mobile flanks of the diapiric bodies. Generally, the contacts of the diapir-like structures with the surrounding rocks are characterized by large thrusts and faults with zones of intense mylonitisation, shearing and tectonic mélanges.

The adjacent Palaeozoic rocks are deflected and curved around the Precambrian core zones forming virgations diverging into opposite vergences (Bard *et al.*, 1973). Around the Münchberg gneiss massif, granite and gneiss-bearing conglomerates in Ordovician, early Late Devonian and Early Carboniferous series already indicate an earlier exposure of the metamorphic basement. Local disconformities, breccias, conglomerates, olistostromes, clastic wedges, rims of shallow-water carbonates and sedimentary gaps are frequently found in the surrounding Palaeozoic series. The Palaeozoic 'Bavarian' facies is restricted to the neighbourhood of the Münchberg gneiss massif and other diapiric complexes in Saxony (Sächsische Zwischengebirge) and reflects the mobility between rising core zones and rapidly subsiding adjacent troughs.

Facies and style of deformation of the adjacent Palaeozoic sedimentary series prove an autochthonous nature and a continuous or repeated diapiric rise of the Precambrian rocks during the Palaeozoic. There are no proofs for giant rootless basement nappes with long lateral displacement (Thiele, 1967; Ries and Shackleton, 1971) or for an emplacement of the basic to ultrabasic rocks due to subduction.

Migration of Troughs

A gradual migration in space and time of Palaeozoic troughs towards the foreland occurred in the external parts away from the crystalline zones and culminated in the Carboniferous molasse foredeeps of the south Portuguese, Cantabrian, Subvariscan and Moravosilesian zones (Fig. 3, Table 1). The pre-flysch series attain thicknesses between 5000 and 10,000 m. In Carboniferous times, narrow foredeeps with an asymmetrical cross-section, filled with thick flysch and molasse sediments, developed in front of the deformed and uplifted internal zones. In contrast to the post-orogenic intramontane molasse basins, marine Namurian flysch or molasse sediments followed Dinantian rocks without interruptions or unconformities in the external foredeeps. The syntectonic flysch and molasse series show a rapid lateral and vertical change in facies and thickness. According to Dvořák (1975) the transition from flysch into molasse sedimentation coincided with the period of most rapid subsidence and most rapid accumulation of clastic material. The total thickness of the external molasse facies attains 5000 m and more, but this amount does not appear in any one place. The transition in depositional environment, from shallow marine to paralic and continental, reflected the final phase of geosynclinal subsidence (Evers, 1967).

TABLE 1. Migration of subsidence and sedimentation from internal to external troughs in North Spain, Rhenish Schiefergebirge and East Sudetes. Thicknesses after Matte (1968), Krebs and Wachendorf (1973) and Dvořák (1975).

	internal ────────────→ external	
North Spain	Westasturian–Leonesian zone	Cantabrian zone
Carboniferous	(only post-orogenic)	6500 m ($f + m$)
Devonian	—	500–2·000 m
Silurian Late Ordovician	3000 m	130– 280 m
Early Ordovician Cambrian	11000 m	1000–2500 m
Rhenish Schiefergebirge	Rhenohercynian zone	Subvariscan zone
Late Carboniferous	several 100 m	4400 m (m)
Early Carboniferous	3000 m (f)	200 m
Late Devonian	1000 m	several 100 m
Middle Devonian	5000 m	
Early Devonian	6000–8000 m	
East Sudetes	Moravosilesian zone (western part)	Moravosilesian zone (eastern part)
Late Carboniferous	—	3000 m (m)
Early Carboniferous		
Late Devonian	8000 m (f)	
Middle Devonian		
Early Devonian		

f = flysch; m = molasse.

In the narrow foredeeps, during the Carboniferous, folding and thrusting gradually migrated in time and space towards the external foreland. The final phase of alpinotype deformation took place in Late Carboniferous times (more or less identical with the Asturian phase). In contrast to the intramontane molasse basins in the internal parts of the Variscan fold belt, the tectonic style of the mobile foredeeps is characterized by folds with outward-directed axial planes, slaty cleavage, large thrusts and superficial décollements. Superficial décollements with horizontal displacements up to 20 km are known from the northern flank of the Stavelot–Venn anticline (Graulich, 1956), Harz Mountains (Schwab, 1974; Lutzens, 1975), and north Spain (de Sitter, 1962; Julivert, 1971; Bard et al., 1973). The metamorphic and igneous rocks of the Lizard, Dodman and Start peninsulas in southwest England probably belong to the northern front of a larger nappe structure. Nappe structures are also known in the Palaeozoic of the Montagne Noire (Rutten, 1969). In some areas the nappes underwent intensive folding after their first emplacement. In the foredeeps, regional metamorphism is generally absent and granitic intrusions are either very scarce or completely absent.

In the eastern Harz Mountains the following succession of gravitational movements can be observed (see Krebs and Wachendorf, 1974):

(1) Devonian pelagic pre-flysch sediments with local slump structures (Blankenburg zone);
(2) Late Devonian to Dinantian autochthonous flysch (Tanne zone);
(3) Late Dinantian olistostromes (Harzgerode zone);
(4) (?) Late Dinantian superficial décollements of flysch and pre-flysch sediments (Südharz and Selke synclines) in their base zones with tectonic mélange.

Stationary Troughs

The troughs in the more internal parts of the Variscan orogen were developed between stabilized Precambrian anticlinal areas and remained nearly stationary during their geosynclinal evolution (Fig. 2). The Palaeozoic series are more or less uninterrupted in the central parts of the troughs, whereas at the rims of the adjoining anticlinal areas, unconformities and sedimentary gaps are significant. To this group belong the troughs in the Armorican Massif (Chateaulin–Laval syncline) and in the Saxothuringian (synclines in Thuringia). The Palaeozoic basin of the Barrandian syncline between the Moldanubian Massif and the Erzgebirge also remained stationary.

In the Amorican Massif the Palaeozoic series are preserved only in narrow elongate synclines, separated by broader anticlinal areas. According to Cogné (1967) the anticlines had rising tendencies, resulting in terrigenous sediments near their' flanks. On the limbs of the Morlaix and Chateaulin–Laval synclines, for example, the Early Devonian rocks transgressed on to Precambrian Brioverian basement (George, 1962).

Locally, deformations (synsedimentary block movements, tilting, uplift, emergence and folding) occurred at the end of the Cambrian, in the earliest Ordovician, at the Silurian–Devonian boundary, in early Late Devonian and in the early Dinantian. Besides the older deformations, the final folding took place in the latest Dinantian or at the boundary between Early and Late Carboniferous (more or less identical with the Sudeten phase).

The post-orogenic molasse sediments rest unconformably, within stationary older synclines (Chateaulin–Laval, Elbe Valley syncline, Doberlug–Kirchhain), on the underlying series. The latest Dinantian and Late Carboniferous–Early Permian intramontane molasse sedimentation is restricted to narrow elongate basins which show an independent sedimentary history for each basin.

Post-orogenic Stage

Parts of the uplifted crystalline core zones, as for example the central German crystalline rise, were inverted at the end of the Dinantian or in the Namurian into graben-like intramontane basins with thick coal-bearing Late Carboniferous–Early Permian series (Saar trough, Saale trough). Other subsequent basins follow anticlinal areas (Malmedy graben near the crest zone of the Stavelot–Venn anticline, Borna–Hainichen basin above the Frankenberg massif) and narrow older Palaeozoic synclines (Chateaulin–Laval syncline in the Armorican Massif, the Elbe Valley syncline in the Saxothuringian zone) or lineaments (Sillon Houiller and other narrow graben structures, bordered by normal faults in the Massif Central).

Eigenfeld and Schwab (1974) pointed out that post-orogenic subsequent volcanism started about 30Ma after the main folding took place, as shown in the table below.

Variscan zones	End of folding	First volcanism
Moldanubian	Dinantian	Westphalian
Saxothuringian	Dinantian	Westphalian
Central German crystalline rise	Namurian	Stephanian
Rhenohercynian	Westphalian	Autunian

The post-orogenic stage was characterized by tension, vertical movements, local folding due to vertical uplifts, granitic intrusions and widespread Late Carboniferous to Early Permian rhyolitic–andesitic–basaltic volcanism. The basins were deformed and folded in variable intensities in the Late Carboniferous or earliest Permian. Thick Early Permian volcanics, associated with red beds and evaporites and locally more than 2000 m thick, also occur in the intracratonic Rotliegende Basin above the Variscan foreland in the subsurface of northern Germany and the southern North Sea (Busch and Kirjuchin, 1972; Gluschko *et al.*, 1975).

Variscan Plate Tectonics?

In recent years several papers have proposed a plate tectonic mechanism for the evolution of the Devonian–Carboniferous Variscan fold belt. In most of the papers the model of a cordilleran-type continental margin is established, which overrode European oceanic crust along a subduction zone. The different models of subduction and continental collision in the Variscan fold belt can be classified into two groups.

(1) The assumption of a *Mid-European ocean* (Saxothuringian oceanic belt, Rheic ocean) with oceanic crust, bordered to the north by the Old Red Sandstone Continent and to the south by a central or south European continental mass. In conformity with the dip of the subduction zones there are three modifications (Fig. 5):

    (a) a northward-dipping subduction zone under the continental margin of the Old Red Sandstone Continent (Johnson, 1973; Riding, 1974);

    (b) a southward-dipping subduction zone under the central or south European continental mass (Laurent, 1972; Mitchell, 1974; Anderson, 1975);

    (c) both northward and southward subduction (Burrett, 1972; Burne, 1973).

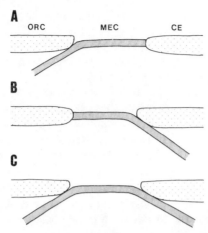

Fig. 5. Different models of subducted oceanic crust in northwestern Europe. ORC: Old
Red Continent; MEC: Mid-European ocean; CE: Central Europe

(2) The assumption of a *Tethyan ocean* (Proto-Mediterranean ocean) between
an African and central or south European continental plate.

There are also two different opinions concerning the dip of a subduction zone to
the north or to the south (Nicolas, 1972; Laurent, 1972; Floyd, 1972; Carvalho,
1972; Soler, 1973; Bard *et al.*, 1973, Riding, 1974). In all these cases the oceanic
lithosphere was progessively consumed during the collision of the above-
mentioned continental masses.

Some of the plate tectonic explanations are restricted to particular areas of
Meso-Europa, but any plate tectonic interpretation should explain the whole
Variscan fold belt. Other papers show only very inaccurate sketches of the
proposed plate boundaries, subduction zones or microcontinents without relation
to geological boundaries seen in the field (for example Burrett, 1972; Riding,
1974; Badham and Halls, 1975). Most papers do not take into consideration the
different vergences of the Variscan fold belt. Theoretically the changing attitudes
of axial planes would imply for the whole *Meso-Europa* at least two or three zones
of consumption with opposite vergences. A northward-dipping subduction zone
under the Old Red Sandstone Continent conflicts with the opposite dip of thrusts
and axial fold planes in the Subvariscan and Rhenohercynian zone (Schroeder,
1973). No model explains the vergence fans in the Iberian Peninsula and in central
Europe or the easternward migration of all Carboniferous foredeeps (for example
Riding, 1974). All plate tectonic explanations deal with the Devonian and Carboniferous history of parts of the Variscan fold belt, but the older metamorphic
and magnetic processes during the early Palaeozoic in the Precambrian core zone
often remain totally disregarded. There is also considerable confusion concerning
the areal distribution of Sudeten and Asturian orogenic movements in *Meso-
Europa* (for example Laurent, 1972; Riding, 1974). In *Meso-Europa*, paired metamorphic belts, widespread true ophiolites, high-pressure metamorphism, mélanges

and linear andesite zones are missing. On the contrary, the Carboniferous granites widely distributed over the whole Variscan fold belt exclude any formation within or above an inclined subduction zone or Benioff zones.

There is no evidence of a Devonian and/or Early Carboniferous oceanic crust south of the Old Red Sandstone Continent.

(1) Zwart (1967) pointed out that in the Variscan fold belt only a few true ophiolite suites occur and ultrabasites are practically absent. The so-called initial magmatism consists mainly of a non-ophiolitic spilite–keratophyre association whereas the typical ophiolitic Alpine ultrabasite–gabbro diabase suite is almost entirely lacking (Ager, 1975; Schermerhorn, 1975). The only exception seems to be the large serpentine–gabbro body of the allochthonous Lizard complex.

(2) There is no uniform or linear greenstone belt in the Variscan fold belt. The keratophyres, spilites and their tuffs occur at different stratigraphical levels, ranging from latest Proterozoic to Early Carboniferous and are widely scattered in different troughs.

(3) The emplacement of the Palaeozoic keratophyre–spilite volcanism was confined to the hinge lines between rising ridges and subsiding basins, caused by deep-seated faulting. Other spilites follow lineaments. This mechanism explains the strongly structure-controlled occurrence of the Palaeozoic keratophyre–spilite association, their different ages in different troughs and their local, but widespread distribution over the Variscan fold belt (Table 2, Fig. 6).

In the Rhenohercynian and Saxothuringian zone, isolated anticlinal structures, which were active during the Devonian as elevations or ridges, were surrounded by effusive and intrusive Devonian spilites and their tuffs. Instructive examples are the East Sauerland anticline and the Hörre–Kellerwald zone in the Rhenish Schiefergebirge (Krebs and Wachendorf, 1974) and the Hirschberg–Gefell anticline in Thuringia (Gräbe et al., 1968). The Precambrian core of the Vosges is also surrounded to the north and south by Devonian and Dinantian basic volcanics (von Eller et al., 1971). Likewise the scarce ultrabasic rock types in *Meso-Europa* are restricted to the rims of the squeezed-out Precambrian basement diapirs. The Palaeozoic series, immediately surrounding the Münchberg gneiss massif, contains Cambrian, Ordovician, Silurian, Devonian and Dinantian submarine basic to acid lava flows and tuffs, whereas in areas further away these volcanics are either very scarce or disappear completely.

(4) It is evident from many regional descriptions that in *Meso-Europa* the submarine pillow basalts, tuffs and associated sills are concordantly interbedded in a marine Palaeozoic sequence. Their maximum thickness only locally attains more than 300 m.

(5) Certainly the basalts are derived from the upper mantle or the lowermost crust and therefore some geochemical similarities may exist with mantle-derived basaltic melts. But geochemically, most of the Devonian and Dinantian spilites in the Rhenohercynian zone represent continental alkaline basalts, quite distinct from oceanic tholeiitic basalts (Herrmann and Wedepohl, 1970; Floyd, 1972).

The entire Variscan fold belt shows a supracontinental evolution because the Palaeozoic sediments always developed on a poly-metamorphic and granitized

TABLE 2. Distribution of pre-orogenic basic (mostly spilitic) volcanics and associated tuffs in Central Europe (for locality numbers see Fig. 6)

Early Carboniferous	Lahn–Dill syncline (3) Kellerwald (4) Western Harz (6) Vosges (23)
Late Devonian	Bergisches Land (1) Lahn–Dill syncline (3) Western Harz (6) margins of the Münchberg gneiss massif (10) East Thuringia (11) and Vogtland (13) Elbe Valley syncline (17) Görlitz Schiefergebirge (20) East Sudetes (25)
Middle Devonian	East Sauerland (2) Lahn–Dill syncline (3) Kellerwald (4) Western Harz (6) Elbingerode complex (7) Wippra zone (8) East Sudetes (25) Vosges (23)
Early Devonian	Hunsrück (5) Lössnitz–Zwönitz syncline (14) Görlitz Schiefergebirge (20)
Silurian	Frankenwald (9) and southwest Vogtland (13) Bardo Mountains (22) Barrandian (24)
Ordovician	margins of the Münchberg gneiss massif (10) margins of the Granulitgebirge (15) Vogtland (13) Elbe Valley syncline (17) Bober–Katzbach Mountains (21) Barrandian (24)
Cambrian	margins of the Münchberg gneiss massif (10) margins of the Granulitgebirge (15) Erzgrebirge (16) Elbe Valley syncline (17) Görlitz Schiefergebirge (20) Bober–Katzbach Mountains (21) Torgau–Doberlug–Göllnitz syncline (19)
Proterozoic	Schwarzburg anticline (12) Erzgebirge (16) Lausitz anticline (18) Barrandian (24)

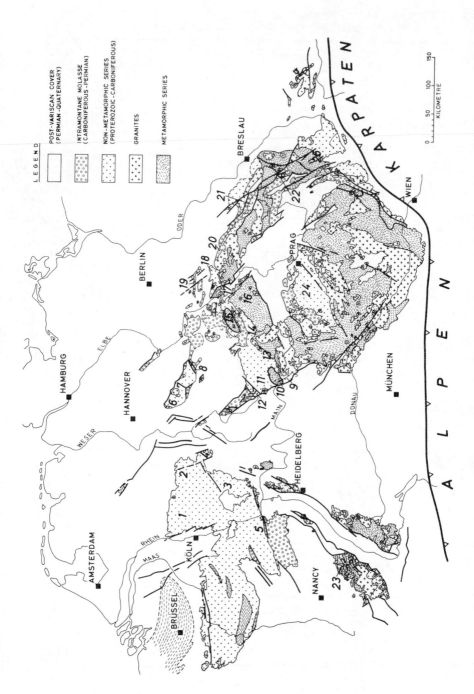

Fig. 6. Geological map of the central European Proterozoic–Palaeozoic (from Krebs and Wachendorf, 1973). The numbers indicate the position of Proterozoic–Early Carboniferous basic 'pre-orogenic' volcanics listed in Table 2

Precambrian basement (Zwart, 1967, 1969; Bard *et al.*, 1973; Krebs and Wachendorf, 1973; Schroeder, 1973). The Palaeozoic non-ophiolitic spilite–keratophyre association has nothing to do with relics of an oceanic crust created at the site of an oceanic ridge. There is no assumption of a former Mid-European oceanic crust, subduction and subsequent Carboniferous continental collision in the Rhenohercynian and Saxothuringian zone (Ager, 1975).

In the Devonian and Early Carboniferous of the south Portuguese zone the palaeogeographic, magmatic, metamorphic and tectonic relationships are absolutely comparable with those of the Rhenohercynian zone. Consequently, the same arguments are valid against the existence of a former *Tethyan* ocean with oceanic crust and associated subduction processes in this belt (Schermerhorn, 1975).

According to Matthews (1974) the so-called 'Variscan Front', running from southern Ireland, southern England, northwestern France and Belgium to the northern Rhenisch Schiefergebirge, refers to different things in different places. There is no reason to suppose that it formed at a uniform deep-seated suture line or plate boundary. The different tectonic style along the 'Variscan Front' results from the influence of older stable massifs or mobile basins at the outermost border of Late Carboniferous folding.

The Vertical Tectonics Model

Although some problems are still unsolved I would like to propose for the evolution of the Variscan *Meso-Europa* a 'fixist' model with primary vertical movements. The mechanism proposed here is based on the models published by Krebs and Wachendorf (1973) and Krebs (1975b).

The foregoing explanations have shown that in the Palaeozoic Variscan fold belt, the metamorphic massifs and large anticlinal areas held a stationary position. From Cambrian to Carboniferous times, subsidence, sedimentation, folding and metamorphism were step-like, migrating from the Precambrian core zones into the adjacent sedimentary troughs (Hoth and Hirschmann, 1970).

As already mentioned, the core zones are characterized by a steady, higher heat flow. These heat flow centres in the crust were developed above ascending asthenospheric diapirs. The rising heat fronts from these mantle plumes or hot spots caused intracrustal regional metamorphism, anatexis, differentiation and formation of granitic melts. The diapiric uprise of lighter melts was due to partial melting, increasing volume and reducing density in contrast to denser overlying metamorphic rocks. The ascent of granitic melts into upper crustal levels caused regional uplift and subsequent erosion. This mechanism explains the local occurrence of conglomerates, clastic rocks and shallow-water carbonates in the neighbourhood of the rising core structures.

The subsidence of corresponding troughs—like the marginal sinks of a rising salt dome—took place synchronously with uplift and erosion of the core zones. Submarine gliding, flysch sedimentation, olistostromes, gravitational folding and superficial décollement structures resulted from the relief between the rising highs and the subsiding troughs. Tectonic mélanges were formed under the weight of overburden along the base of large imbricate thrust sheets and nappes, for example Meneage crush zone–Lizard complex and sheared rocks at the base of the Südharz 'Syncline'–Harz Mountains. Basic melts from the upper mantle or the lowermost crust ascended on the mobile hinge lines between zones of different vertical movements. Local vergence fans marked the gravitational spreading of the vertically ascending mobile masses as opposed to the subsiding basins.

Generally, the primary tectogenesis with vertical uprise due to rising heat flow fronts, with steep vertical gradients, induced a secondary tectogenesis with centripetally directed allochthonous complexes, thrusts and folds.

Finally, the Palaeozoic fold belt of Variscan *Meso-Europa* is not a result of continental collision, including consumption of oceanic crust, but folding and thrusting are due to the vertical rising and lateral spreading of hot mobile core zones. Metamorphism, granitization, volcanism and deformation are the consequence of repeatedly rising asthenospheric plumes or diapirs which remained stationary and spread outwards during the Palaeozoic and which were widely distributed over the whole *Meso-European* fold belt.

Although the model proposed here seems to be a simpler method of explaining the geosynclinal evolution of *Meso-Europa* than the application of plate tectonics, several specific problems remain unsolved and many questions are still open to future research.

References

Ager, D. V. (1975). 'The geological evolution of Europe', *Proc. Geol. Ass.*, **86**, 127–54
Anderson, T. A. (1975). 'Carboniferous subduction complex in the Harz Mountains, Germany', *Bull. geol. Soc. Am.*, **86**, 77–82
Aubouin, J. (1965). *Geosynclines: Developments in Geotectonics*, Vol. 1, Elsevier, Amsterdam, London, New York, 335 p.
Badham, J. P. N. and Halls, C. (1975). 'Microplate tectonics, oblique collisions, and evolution of the Hercynian orogenic systems, *Geology*, **3**, 373–6
Bailey, E. B. (1928). 'The Palaeozoic mountain systems of Europe and America', *Rep. Br. Ass. Adv. Sci.,* sec. C, 57–76
Bard, J.-P., Capdevila, R., Matte, P. and Ribero, A. (1973). 'Geotectonic model for the Iberian Variscan orogen, *Nature phys. Sci.,* **241**, 50–2
Bernstein, K.-H., Blüher, H.-J., Bolduan, H., Brause, H., Douffet, H., Hirschmann, G., Hoth, K., Lorenz, W., Mibus, P., Mucke, C. and Scheumann, H. (1973). Erläuterung zur geologischen Übersichtskarte der Bezirke Dresden, Karl-Marx-Stadt und Leipzig 1: 400,000: VEB Geologische Forschung und Erkundung Halle, 78
Brause, H. (1970). 'Variszischer Bau und "Mitteldeutsche Kristallinzone"', *Geologie*, **19**, 281–92
Bubnoff, S. V. (1930). *Geologie von Europa*, Vol. 2, Das außeralpine Westeuropa, Teil 1: Kaledoniden und Varisciden: Borntraeger, Berlin, 690 p.
Burne, R. V. (1973). 'Palaeogeography of South West England and Hercynian continental collision', *Nature phys. Sci.,* **241**, 129–31
Burrett, C. F. (1972). 'Plate tectonics and the Hercynian orogeny', *Nature*, **239**, 155–7
Busch, W. A. and Kirjuchin, L. G. (1972). 'Über die Verbreitung subsequenter Effusiva des Jungpaläozoikums in Mitteleuropa', *Z. angew. Geol.*, **18**, 323–8
Carvalho, D. de (1972). 'The metallogenetic consequences of plate tectonics and the Upper Palaeozoic evolution of southern Portugal', *Estud. Not. Trab. Serv. Fom. Min.*, **20**, 297–320

Cogné, J. (1967). 'Les grand cisaillements Hercyniens dans le Massif armoricain, et les phénomènes de granitisation', in Schaer, J. P. (ed.), *Colloque Etages Tectoniques*, Baconniere, Neuchâtel, 179–92

Dvořák, J. (1975). 'Interrelationship between the sedimentation rate and the subsidence during the flysch and molasse stage of the Variscan geosyncline in Moravia (Sudeticum)', *N. Jb. Geol. Paläont.*, Mh., 339–42

Dvořák, J. and Paproth, E. (1969). 'Uber die Position und die Tektogenese des Rhenoherzynikums und des Sudetikums in den mitteleuropäischen Varisziden', *N. Jb. Geol. Paläont.*, Mh., 65–88

Eigenfeld, F. and Schwab, M. (1974). 'Zur geotektonischen Stellung des permosilesischen subsequenten Vulkanismus in Mitteleuropa', *Z. geol. Wiss.*, **2**, 115–37

Eller, J.-P. v., Hameurt, J. and Ruhland, M. (1971). 'Evolutions métamorphiques et ignées dans les Vosges', *Soc. Géol. France C. R.*, **6**, 296–306

Evers, H. J. (1967). 'Geology of the Leonides between the Bernesga and Porma rivers, Cantabrian Mountains, NW Spain', *Leidse Geol. Med.*, **41**, 83–151

Floyd, P. A. (1972). 'Geochemistry, origin and tectonic environment of the basic and acidic rocks of Cornubia, England', *Proc. Geol. Ass.*, **83**, 385–403

Gaertner, H. R. v. (1969). 'Zur tektonischen und magnetischen Entwicklung der Kratone', *Beih. Geol. Jb.*, **80**, 117–45

George, T. N. (1962). 'Devonian and Carboniferous foundations of the Variscides in north-west Europe', in Coe, K. (ed.), *Some Aspects of the Variscan Fold Belt*, Manchester Univ. Press, 19–47

Gluschko, W. W., Hetzer, H., Katzung, G., Dikenschtejn, G. Ch., Solowjew, B. A. and Tshchernyschew, S. M. (1975). 'Grundzüge des geologischen Baus und der Gasführung des Rotliegenden in der Mitteleuropäischen Senke', *Z. angew. Geol.*, **21**, 253–62

Gräbe, R., Schlegel, G. and Wucher, K. (1968). 'Environment and palaeogeography of the Devonian in the area of the Berga anticline, Thuringia, Germany', Internat. Symp. Devonian System. Calgary 1967, *Alberta Soc. Petrol. Geol.*, **2**, 1283–96

Graulich, J. M. (1956). 'La situation géologique du sondage de Soumagne (Pays de Herve)', *Bull. Soc. belg. Géol. Paléont. Hydrol.*, **65**, 339–45

Hempel, G. (1974). 'Variscische Tektogenese', in Hoppe, W. and Seidel, G. (eds.), *Geologie von Thüringen*, Haack, Gotha/Leipzig, 289–335

Herrman, A. G. and Wedepohl, K. H. (1970). 'Untersuchungen an spilitischen Gesteinen der variskischen Geosynklinale in Nordwest-deutschland', *Contrib. Mineral. Petrol.*, **29**, 255–74

Hoth, H. and Hirschmann, G. (1970). 'Das Jungpräkambrium im Bereich der Varisziden und Kaledoniden West- und Nordeuropas und seine Beziehungen zu den paläozoischen Entwicklungsetappen', *Ber. dt. Ges. geol. Wiss.*, A, **15**, 379–424

House, M. R. (1975). 'Facies and time in Devonian tropical areas', *Proc. Yorkshire geol. Soc.*, **40**, 233–88

Johnson, G. A. L. (1973). 'Closing of the Carboniferous sea in western Europe', in Tarling, D. K. and Runcorn, S. K. (eds.), *Implications of Continental Drift to the Earth Sciences*, Vol. 2, Academic Press, 845–50

Julivert, M. (1971). 'Décollement tectonics in the Hercynian cordillera of northwest Spain', *Am. J. Sci.*, **270**, 1–29

Kossmat, F. (1927). 'Gliederung des varistischen Gebirgsbaues', *Abh. sächs. geol. Landesamt*, **1**, 39 pp.

Krebs, W. (1975a). 'Geologische Aspekte der Tiefenexploration im Paläozoikum Norddeutschlands und der südlichen Nordsee', *Erdoel-Erdgas Z.*, **91**, 277–84

Krebs, W. (1975b). 'Evolution of southwest Pacific island arc-trench systems and mountain belts: plate tectonics or global gravity tectonics?' *Bull. Am. Ass. Petrol, Geol.*, **59**, 1639–66

Krebs, W. and Wachendorf, H. (1973). 'Proterozoic–Paleozoic geosynclinal and orogenic evolution of Central Europe', *Bull. geol. Soc. Am.*, **84**, 2611–29

Krebs, W. and Wachendorf, H. (1974). 'Faltungskerne im mitteleuropäischen Grundgebirge–Abbilder eines orogenen Diapirismus', *N. Jb. Geol. Paläont., Abh.*, **147**, 30–60

Laurent, R. (1972). 'The Hercynides of South Europe, a model', *24th Internat. geol. Congr.*, Sect. 3, 363–70

Lotze, F. (1945). 'Zur Gliederung der Varisziden der Iberischen Meseta, *Geotekt. Forsch.*, **6**, 78–92

Lotze, F. (1963). 'Die variszischen Gebirgszusammenhänge im westlichen Europa', *Geol. Ann. Mus. Geol. Bologne, ser. 2*, 31

Lutzens, H. (1975). 'Ein Beitrag zur Geologie des Unterharzes–Metamorphe Zone, Südharz- und Selkemulde', *Z. geol. Wiss.*, **3**, 267–99

Matte, P. (1968). 'La structure de la virgation hercyniene de Galice (Espagne)', *Géol. Alpine*, **44**, 157–280

Matthews, S. C. (1974). 'Exmoor thrust? Variscan Front?', *Proc. Ussher Soc.*, **3**, 82–94

Meinel, G (1974). 'Magmatismus und Metamorphose', in Hoppe, W. and Seidel, G. (eds.), *Geologie von Thüringen*, Haack, Gotha/Leipzig, 335–52

Mitchell, A. H. G. (1974). 'Southwest England granites: magmatism and tin mineralization in a post-collision tectonic setting', *Trans. Inst. Min. Metall., sect. B*, **83**, B95–B97

Morawski, T. (1973). 'The Sowie Góry area and its petrological problems', in Smulikowski, K. (ed.), *Revue des Problèmes Géologiques des Zones Profondes de l'Écorce Terrestre en Basse Silésie*, Inst. Sci. Géol. Acad. Polonaise Sci., 44–58

Nicolas, A. (1972). 'Was the Hercynian orogenic belt of Europe of the Andean type?', *Nature*, **236**, 221–3

Paproth, E. and Wolf, M. (1973). 'Zur paläogeographischen Deutung der Inkohlung im Devon und Karbon des nördlichen Rheinischen Schiefergebirges', *N. Jb. Geol. Paläont., Mh.*, 469–93

Riding, R. (1974). 'Model of the Hercynian foldbelt', *Earth Planet., Sci. Letters*, **24**, 125–35

Ries, A. C. and Shackleton, R. M. (1971). 'Catazonal complexes of North-West Spain and North Portugal, remnants of a Hercynian thrust plate', *Nature phys. Sci.*, **234**, 65–8, 79

Rutten, M. G. (1969). *The Geology of Western Europe*, Elsevier, Amsterdam, 520 p.

Schermerhorn, L. J. G. (1975). 'Spilites, regional metamorphism and subduction in the Iberian Pyrite Belt: some comments', *Geol. Mijnbouw*, **54**, 23–35

Schmidt, K. and Franke, D. (1975). 'Stand und Probleme der Karbonforschung in der Deutschen Demokratischen Republik. Teil I: Unterkarbon', *Z. geol. Wiss.*, **3**, 819–49

Schoenenberg, R. (1971). *Einführung in die Geologie Europas*, Rombach, Freiburg, 300 p.

Schoenenberg, R. (1973). 'Zur Frage der Verbindung von Sudetikum und ostalpinem Variszikun', *Veröff. Zentr. Inst. Physik Erde*, **14**, pt. 2, 437–50

Schroeder, E. (1973). 'Probleme tektonischer Untersuchungen im Orogen, speziell in den Varisziden', *Veröff. Zentr. Inst. Physik Erde*, **14**, pt. 2, 273–302

Schwab, M. (1974). 'Harz-verkehrt gestapelt. Neue Theorien zum Gebirgsbau des Harzes', *Wissenschaft u. Fortschritt*, **24**, 85–9, 140–5

Sitter, L. U. de (1962). 'The Hercynian orogenes in northern Spain', in Coe, K. (ed.), *Some Aspects of the Variscan Fold Belt*, Manchester Univ. Press, 1–18

Soler, E. (1973). 'L'association spilites–quartz kératophyres du sud-ouest de la péninsule ibérique', *Geol. Mijnbouw*, **52**, 277–87

Stille, H. (1924). *Grundfragen der Vergleichenden Tektonik*, Borntraeger, Berlin, 443 p.

Stille, H. (1951). 'Das mitteleuropäische variszische Grundgebirge im Bilde des gesamteuropäischen', *Beih. Geol. Jb.*, **2**, 138 p.

Thiele, O. (1967). 'Die Munchberger Gneismasse als Zeugnis für den Deckenbau der Varisziden', *Nitt. geol. Ges. Wein*, **59**, 219–29

Weber, K. (1972). 'Kristallinität des Illits in Tonschiefern und andere Kriterien schwacher Metamorphose im nordöstlichen Rheinischen Schiefergebirge', *N. Jb. Geol. Paläont., Abh.*, **141**, 333–63

Wolf, M. (1972). 'Beziehungen zwischen Inkohlung und Geotektonik im nördlichen Rheinischen Schiefergebirge', *N. Jb. Geol. Paläont., Abh.*, **141**, 222–57

Ziegler, P. A. (1973). 'Geologic evolution of North Sea and its tectonic framework', *Bull. Am. Ass. Petrol. Geol.*, **59**, 1073–97

Zwart, H. J. (1967). 'The duality of orogenic belts', *Geol. Mijnbouw*, **46**, 283–309

Zwart, H. J. (1969). 'Metamorphic facies series in the European orogenic belts and their bearing on the causes of orogeny', *Geol. Ass. Canada, spec. pap.*, **5**, 7–16

Neo-Europa

That part of the continent directly
affected by the main Alpine tectonism

ALPINE TECTONICS AND PLATE TECTONICS: THOUGHTS ABOUT THE EASTERN MEDITERRANEAN

JEAN AUBOUIN

Département de Géologie Structurale, Université Pierre et Marie Curie, 75230 Paris, France

Abstract

Aubouin, J. (1976). 'Alpine tectonics and plate tectonics: thoughts about the eastern Mediterranean', in Ager, D. V. and Brooks, M. (eds.), *Europe from Crust to Core*, Wiley, London.

The existence of superimposed tectonic events in the Mediterranean area is emphasized; retrotectonic analysis leads to the recognition of the following tectonic sequence:

(5) Neotectonics or Mediterranean tectonics;
(4) Alpine late tectonics or deformation of continental margins after their collision;
(3) Alpine tectonics or mesogean closing and collision of continental margins;
(2) Alpine palaeotectonics or Pacific stage of the mesogean palaeo-ocean;
(1) Alpine pretectonic framework or Atlantic stage of the mesogean palaeo-ocean.

The main structural periods are: Triassic with Atlantic-type opening and spreading of a mesogean palaeo-ocean; Jurassic–Cretaceous boundary with the birth of a compressive Pacific-type situation (Upper Jurassic revolution); Miocene/Pliocene boundary with the birth (mainly by spreading) of the present Mediterranean Sea (Pliocene revolution). During the Cretaceous (to Eocene) Pacific stage the Eurasian margin was of Andean type, the Afroarabian margin of Pacific island arc type, with an asymmetry similar to that of the present Pacific one. During the collision stage, the Afroarabian plate was the main foreland which, being autochthonous, passes under the allochthonous palaeo-oceanic formations. The chronology changes spacially, being earlier towards the east and later towards the west. The structural situation changes also: in the eastern Alps the main foreland is the Eurasian plate; thus transverse Dinaric structures are of special interest.

Many attempts have been made to interpret Alpine perimediterranean tectonics in terms of plate tectonics but, generally, neglecting the superimposed (overprinted) tectonics during the alpine cycle. Thus, the seismic data are used to interpret the edifice of nappes built during the Mesozoic and Cenozoic and the present continental and oceanic crust distribution is used as a guide to the past distribution.

In this paper I shall emphasize the sequence of superimposed tectonic events in the alpine cycle (Fig. 1): palaeotectonics, the main thrusting, late tectonics (deformation of the thrust sheets), neotectonics (the last structures, which are still active; see Aubouin, 1973).

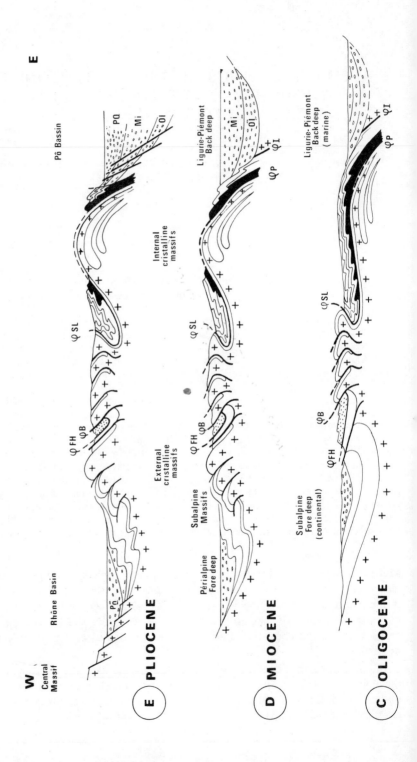

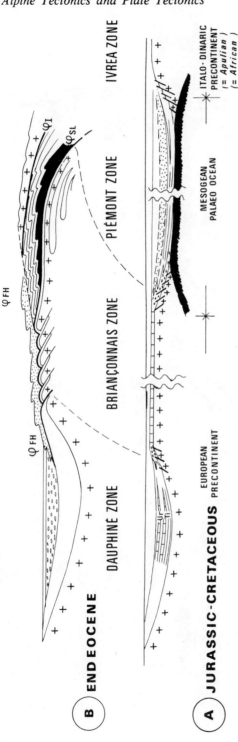

Fig. 1. Very simplified sketch of the evolution of the western Alps. FH, Helminthoid flysch; Mi, Miocene; Ol, Oligocene; PQ, Plio–Quaternary; SL, Schistes Lustrés; Ti, Upper Jurassic (Tithonian); Ur, Urgonian; black, ophiolites; φ, Main thrusts; φB, Briançonnais thrust; φFH, Helminthoid flysch thrust; φI, Ivrea thrust; φSL, Schistes Lustrés thrust. *Note*: (i) the succession, tectonic thrusts (B), late tectonic basement folding (C, D), neotectonic faulting (E); (ii) the possible palaeo-oceanic significance of the Piedmont trough during Jurassic (and partly Cretaceous) times, between two 'Atlantic-type' margins; (iii) the infrastructural deformation of the European continental margins (e.g. Mont Rose nappe, Simplon–Tessin nappes, etc.) deformed by the late tectonic basement folding of the internal crystalline massifs.

The sketch is very incomplete; among other things the palaeotectonic structures formed during Cretaceous times are lacking (see Fig. 5).

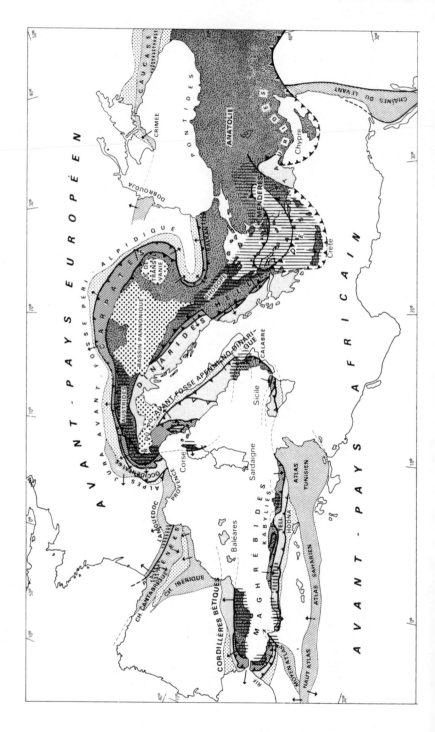

Fig. 2. Tectonic sketch map of the perimediterranean chains (after Aubouin and Durand-Delga, 1971). *Note:* (i) the (relative) independence of the Mediterranean Sea with reference to the alpine framework (see Fig. 3); the situation of the eastern Mediterranean window of the Afroarabian plate (Attica, Cyclades, Menderes massif): these are not individual microplates (see Fig. 4). In this connection, the representation of Eastern Anatolia is not significant: there are other crystalline massifs which probably have the same significance as the windows of the Afroarabian foreland; (ii) the importance of transverse structures such as the Scutari–Pec line between the Dinarides and Hellenides (see Fig. 6) or the Isparta line between the Hellenides and the Taurides

Neotectonics of Mediterranean Tectonics

The Mediterranean region is characterized by a network of active faults with two main directions: NW–SE and NE–SW. This network came into existence at the end of the Miocene and has experienced two main episodes: one at the Miocene–Pliocene boundary (so-called Pontian movements) another at the Pliocene–Quaternary boundary (so-called Villafranchian movements). This network cuts sharply across earlier alpine structures, being transverse, longitudinal and oblique with regard to them: it does not inherit earlier alpine tectonic trends. It determines the Mediterranean Sea in its present position and mountain chains and intervening intramontane plains are block-faulted horst and graben, respectively.

This faulting generates the *Mediterranean framework*, quite different from the alpine framework; moreover the Mediterranean Sea itself stands out of the alpine domain, cutting its foreland, especially in the southeast (Fig. 2).

Thus the Mediterranean Sea was born after a *Pliocene revolution* (Bourcart, 1962), or rather a *Messinian revolution*, since the large lagoon of this age had the shape of the future Mediterranean.

In many papers the Mediterranean is related to Miocene or Oligocene marine basins after either the Messinian evaporite episode or a stratigraphic discontinuity. With regard to alpine tectonics these basins are only of late-tectonic meaning; they have the situation of molasse troughs, already distinct from the alpine palaeogeographic units, but without any relation to the Plio–Quaternary Mediterranean Sea.

On the whole, neotectonic features are extensional, with an important strike-slip component. In detail, they are more complex (see Mercier *et al.*, 1972, 1975; Angelier, 1973) in time (thus, between the two main extensional episodes in the Aegean domain of Pontian and Villafranchian age some compressional movements occurred) and in space (thus, on the Aegean arc periphery, compressional movements carry on during the extensional phase in the Aegean Sea itself). The same observation can be made about the Tyrrhenian arc periphery. However, extensional structures are very much larger than compressional ones. The magnitude of vertical components is of the order of kilometres for the former and metres for the latter.

The Mediterranean arcs are the products of this neotectonic episode. For instance, one frequently thinks that the Oligo–Miocene magmatism (andesite, granodiorite) commonly observed in the eastern Mediterranean area is due to periaegean subduction (McKenzie, 1970, 1972; Caputo *et al.*, 1970; Ninkovitch and Hays, 1972). But the very wide distribution of this magmatism makes this direct relation impossible in space and in time. Only the Cycladic magmatism, of Plio–Quaternary age, is linked to Aegean arc subduction defining an internal volcanic arc from the Saronic Gulf to the Turkish coast by Poros, Methana, Milos, Santorini (Thira), Nisiros. If the Oligo–Miocene magmatism is due to subduction, it takes place along another line related to alpine tectonics: the Vardar zone, for instance.

Mediterranean seismic belts have only a neotectonic significance: for instance, the plates described in the eastern Mediterranean (McKenzie, 1970, 1972; Fig. 3) like the Aegean plate and the Turkish plate are *neoplates*. The northwestern limit of the Aegean plate cuts perpendicularly across all the alpine structures up to those of Late Miocene age; moreover this northwestern limit is a transverse fault zone of Pliocene age ('couloir de Karpenision' in continental Greece—see Aubouin and Guernet, 1963) with a part of new oceanic crust in the Macedonian trough of extensional character (Le Pichon *et al.*, 1973). So the new subduction at the Aegean periphery and the new extensional formation of the Aegean Sea are two linked phenomena, like the classical model of island arc/marginal sea.

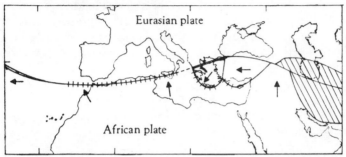

Fig. 3. Mediterranean neoplates (from McKenzie, 1970). *Note* the independence of the plate limits with regard to the alpine framework, especially in the eastern Mediterranean domain (see Fig. 2)

The present Mediterranean framework is thus a neotectonic framework, related more to the future than to the past. Mediterranean plates are *neoplates*; Mediterranean oceanic zones are *neo-oceans* and so on. Likewise the main faults are *neofaults*; for example the North Anatolian fault whose relations with the alpine structures are of the same type as the relations between the San Andreas fault and Californian nappes: the faults are superimposed. The Eurasian–Afroarabian suture of alpine age is the North Anatolian ophiolitic zone, clearly cut by the North Anatolian fault (Bergougnan, 1975; Fourquin, 1975).

Neotectonics characterizes the so-called *postgeosynclinal period* of alpine evolution (Aubouin, 1961, 1965); the name implies the complete independence of this new period.

Alpine Late Tectonics or Deformation of Continental Margins After their Collision

Apart from neotectonic features, all the perimediterranean chains appear as associations of large anticlinal basement folding and large synclinal molasse troughs, the whole being of Oligo–Miocene age. These anticlinal structures of large radius affect the previous thrusts and the corresponding autochthon; they determine that the highest points in the topography (for instance, the Mont Blanc Massif in the Alps) are emplaced in the lowest structural units (basement of the autochthonous external zone; see Fig. 1).

The style is basement folding, involving continental crust. The basement may crop out, as in the external and internal crystalline massifs of the Alps, or not, as with the Olympus marble sedimentary cover of the Gavrovo zone belonging to Apulian–African continental margin. Thus, the nappes are deformed together with the continental border over which they are thrust and which can be remobilized in the infrastructural domain with more or less metamorphism (see Fig. 1).

This implies that the continental margins of Europe and Africa were in contact during the Late Eocene tectonic phase, which was the main alpine orogenic phase. As a consequence, the whole alpine chain is quite allochthonous, except the narrow suture (ophiolitic) zone which is the scar of the Mesogean palaeo-ocean, for example, the Vardar zone of the Dinarides. This must be qualified for the external zones which are part of the continental margins, since they are more or less autochthonous (relative to the completely allochthonous internal nappes of palaeo-oceanic origin).

The basement folds give way to *décollement* tectonics and some cover thrusts are contemporaneous: in the external zones the cover nappes are thrust during the Miocene. Moreover the basement folding carries on continuously during Oligocene and Miocene as a consequence of the progressive blocking of the collision.

The so-called *late geosynclinal period* (Aubouin, 1961, 1965) is characterized by its molasse troughs and its granodioritic–andesitic magmatism resulting, perhaps, from the remobilization of the lower part of the continental crust during the formation of a mountain root.

Alpine Tectonics, Mesogean Closing and Collision of Continental Margins

The Alpine tectonics proper consist primarily of a tremendous piling up of nappes, and a corresponding high pressure–low temperature metamorphism. This Late Eocene phase is the end of a period of deformation continuous since the end of the Cretaceous, with a characteristic polarity in each chain: deformation is older in the internal zones, younger in the external ones. This, in agreement with the tectonic vergence, permits the clear individualization of each mountain range: the Alps, Apennines, Dinarides, etc.

The ophiolitic nappes, which are the main elements of alpine structures, originate from the Mesogean palaeo-oceanic crust (De Roever, 1957; Dercourt, 1970, 1972; Dewey and Bird, 1970; Moores, 1969, etc.). Thus the alpine chains are interpreted as being derived from the closing of the Mesogean palaeo-ocean by collision of the Eurasian and the Afroarabian continental margins (Le Pichon, 1968; Morgan, 1968; Dewey and Bird, 1970, etc.), a new expression of the shortening between Europe and Africa proposed long ago by Argand. The main oceanic scar runs from the Vardar zone in the Balkan peninsula to the North Anatolian ophiolitic zone and South Caucasus; elsewhere it is hidden by a Recent sedimentary cover (Pannonic basin, Mediterranean Sea) or is conjectural (the tonalitic line?, Insubric line?, Sestri–Voltaggio line?).

It is thus possible to give an interpretation in the light of plate tectonics based, on the one hand, on the alpine structures and, on the other hand, on movements of the Eurasian and Afroarabian plates, deduced from Atlantic magnetic anomalies (Dewey *et al.*, 1973). But was the Mesogean palaeo-ocean as simple as the present Atlantic or more complex with a number of microplates between Eurasia and Africa (quite different from the Mediterranean neoplates of neotectonic nature)? The geophysical arguments fit both hypotheses.

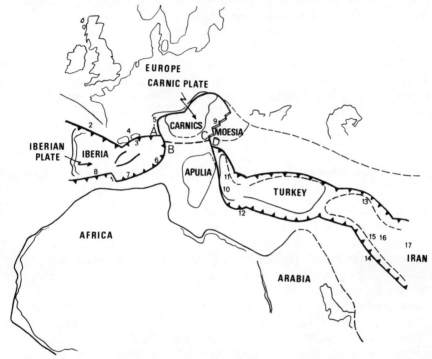

Fig. 4. Mediterranean palaeoplates during the Eocene stage (from Dewey *et al.*, 1973). *Note* that the eastern Mediterranean plates are not individual plates, but metamorphic sequences of Afroarabian continental margin in the position of tectonic windows (see Fig. 2)

Owing to the Alpine complexity, especially the numerous outcrops of metamorphic rocks, various numbers of microplates have been proposed (Smith, 1971; Dewey *et al.*, 1973, etc.; Fig. 4). Thus we have the Adriatic (Apulian) microplates, formed of the Italo–Dinaric system (with its centripetal symmetry); but this is simply an African promontory, clearly moved to the NNW by longitudinal motion (Aubouin, 1961, 1965). Similarly the Aegean, Turkish and Iranian microplates are distinguished in an intermediate position between the two main plates in the eastern Mediterranean area; they correspond to the Attic–Cycladic metamorphic outcrops and, more generally, to the central Anatolian metamorphic massifs; but they are only outcrops of the Afroarabian continental margins, metamorphosed in infrastructural position, and appearing as tectonic windows: these are the Olympus (Godfriaux, 1962), Attic–Cyclad massif (Katsikatsos, 1971), Menderes massif (Durr, 1975) and the main Anatolian crystalline massifs (Argyriadis and Ricou, 1975). They are not microplates, but late anticlinal outcrops of Afroarabian continental margin, under the Mesogean nappes (see above and Fig. 2).

Not every metamorphic outcrop underlying a nappe system is necessarily an isolated continental plate. It may be the infrastructural part of the tectonic edifice, especially of the autochthonous continental margin. Metamorphic outcrops and basement outcrops are still too often confused. As for the isolation or not, of such microplates, the biostratigraphic arguments are of primary interest: the so-called Apulian (Italo–Dinaric) plate belongs to Africa; and Asia Minor windows show Gondwanan faunas while the North Anatolian zone presents Eurasian faunas (Enay, 1972). It seems that, in the alpine framework of the Mediterranean area, complexity is tectonic and not palaeogeographic. Perhaps there were no microplates at all between the Eurasian and Afroarabian plates.

Alpine Palaeotectonics or the Pacific Stage of the Mesogean Palaeo-ocean

Before the main thrusting due to collision, many important tectonic phases are known. Disregarded for a long time, except in the Carpathians and the eastern Alps where Late Cretaceous phases have been described, they are now a subject of great interest. The Cretaceous appears to be a period of constant orogeny, with strong episodes at the Jurassic–Cretaceous boundary, the Mid-Cretaceous, and during the Late Cretaceous.

These palaeotectonic events give birth to nappes involving ophiolitic massifs generated in the oceanic crust, and associated sedimentary formations. These nappes seem to belong to island arcs of western Pacific type; and their thrusting is independent of (i.e. older than) the later collision. This is very clear in the Dinarides, where the first thrusting took place at the Jurassic–Cretaceous boundary (Fig. 5); and the *periarabian ophiolitic crescent* (Ricou, 1972) of Late Cretaceous age is very similar to the *periaustralian ophiolitic crown* of Palaeogene age, for instance, running across New Caledonia and New Guinea and beyond (Aubouin, Mattauer and Allegre, in press).

In the eastern Mediterranean area, the main foreland was the Afroarabian plate towards which are directed the main thrust movements of the palaeotectonic and tectonic episodes. In contrast, the Eurasiatic foreland is less deformed, but is the site of a very strong magmatism of andesitic and granodioritic type. This disposition is similar to the present Pacific asymmetry with a western island arc system with ophiolitic nappes, and an eastern cordilleran system with extensive granodioritic–andesitic magmatism. Thus, during the Cretaceous, the Mesogean palaeo-ocean was of Pacific type with an Afroarabian margin fringed with island arcs, and a Eurasiatic margin of Andean type, at least in the eastern Mediterranean area.

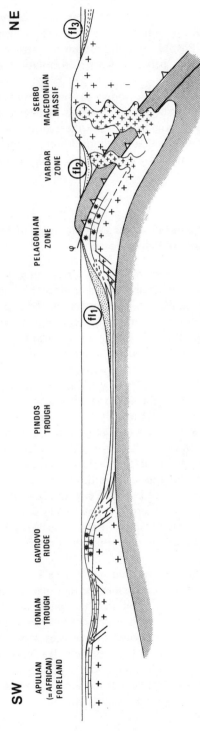

Fig. 5. Possible state of the Dinarides at the Jurassic–Cretaceous boundary. *Note:* (i) the Atlantic-type border of the Apulian (Arabian) margin; (ii) the first collision between the island arc of western Pacific type (Pelagonian zone) and the Cordilleran margin of eastern Pacific type (Serbo–macedonian massif); (iii) the Pindus trough as a marginal sea of western Pacific type; (iv) the granodiorites on the Serbo–macedonian (Cordilleran) margin; (v) the three detrital sequences: fl_1, of the Bosnian flysch; fl_2, of the Maglaj flysch; fl_3, of the Balkan flysch. If correct, the Mesogean palaeo-ocean was in the Vardar zone before the collision of the two margins: i.e. (i) the African margin, of western Pacific type, with the Pindus marginal sea and the Pelagonian island arc with the first ophiolitic palaeothrust; and (ii) the Eurasian margin of eastern Pacific type with the Serbo–macedonian and (Rhodope) massifs as an Andean Cordillera

In the western Mediterranean, things are not so clear; asymmetry is not well marked, and is even of opposite direction in the eastern Alps, which marks the African thrusting over Europe. Possibly this disposition is restricted to the extremity of the Italo–Dinaric system, between two strike-slip zones (the Vardar zone and the Sestri–Voltaggio zone).

The flysch of the so-called *orogenetic stage* of the geosynclinal period (Aubouin, 1961, 1965) or *filling stage* begins, in the whole Mediterranean area, at the Jurassic–Cretaceous boundary, originating from the erosion of island arcs and not of the foreland. Andesites and granodiorites were abundant on the Eurasiatic margin which took an Andean aspect in relation to the main subduction towards the Eurasiatic continent.

Alpine Pretectonic Framework or the Atlantic Stage of the Mesogean Palaeo-ocean

Before the compressional revolution of the Jurassic–Cretaceous boundary, the Mesogean palaeo-ocean had been opening and spreading since the Triassic: it was an Atlantic-type ocean with two similar margins (except for the biostratigraphic difference). Some transverse tectonic features may represent palaeotransform faults of this Mesogean palaeo-ocean; such are the Dinaric transverse structures which cut the ophiolitic nappes of palaeo-oceanic origin but not (or at least to a lesser degree) the Apulian (African) continental margin (Fig. 6; Aubouin and Dercourt, 1975).

This disposition was born during the Triassic in the western Mediterranean area where the Mesogean palaeo-ocean was created by the rifting of the Hercynian domain: it is the *recreated western Mesogea*, (or *Tethys*), with a characteristic sequence of facies: continental detritus of the Early Triassic, evaporites of the Middle–Late Triassic, oceanic facies of the Late Triassic. This sequence recalls the birth of the Atlantic Ocean in other times. Towards the eastern Mediterranean area the phenomenon becomes older, and from Permian times the facies are oceanic. Beyond that, towards the east there is no more Variscan orogeny and the Mesogean ocean is directly inherited from the Palaeozoic (Argyriadis, 1975): it is the *eastern permanent Mesogea* in opposition to the recreated western Mesogea.

The so-called *static stage* of the geosynclinal period (Aubouin, 1961, 1965) may be referred to as the *vacuity stage*, expressing the fact that sedimentation is very slow in opposition to the flysch sedimentation of the filling stage (see above). The key sediments are radiolarites in the palaeo-oceanic deeps and nodular limestones of *ammonitico rosso* type on the saddle; sometimes with detrital facies which, in opposition to flysch, originated in the forelands. It is easy to compare classical geosynclinal terminology with plate terminology during this stage: continental margins are miogeosynclinal zones, oceans are eugeosynclinal zones (Drake, Ewing and Sutton, 1959; Kay, 1951; Aubouin, 1961, 1965; Dietz and Holden, 1966, etc.).

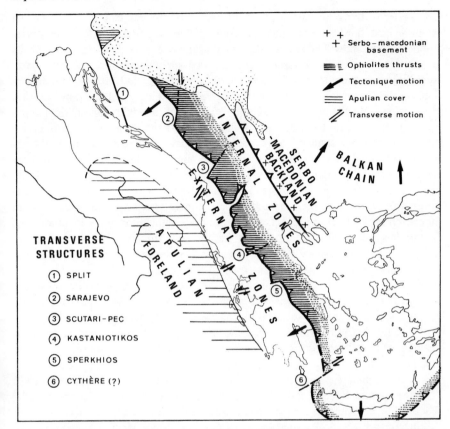

Fig. 6. Dinaric transverse structures as inherited from palaeotransform faults (from Aubouin and Dercourt, 1975). *Note* the ophiolitic thrust front, shifted by transverse structures which have no effect on the external zones (= Apulian–African continental margin). Apart from the Split transverse zone, this was perhaps the most important in the Mediterranean domain (or the Scutari–Pec line?—see Fig. 2). As ophiolites are part of the palaeo-oceanic crust, the transverse zones may be inherited from the palaeotransform faults, later reworked

Conclusions

In so short a paper it is clearly impossible to cover all the Mediterranean and Alpine problems. Among others, we have not examined the Italo–Dinaric longitudinal motion towards the NNW during Cretaceous and Early Tertiary times, responsible for the eastern Alps genesis. Nor have we considered the Corsican–Sardinian rotation during Mid-Tertiary, which complicates the succession of tectonics, late tectonics and neotectonics in the western Mediterranean. Before the Plio–Quaternary Mediterranean Sea, there was an older western Mediterranean during the Oligocene and Miocene, for example. In summary, the retrotectonic method delineates the following sequence of superimposed tectonic events in the Mediterranean area. During the Triassic: Atlantic-type opening of the western Mesogean palaeo-ocean, by rifting of the Variscan domain, starting from an eastern Mesogean, inherited from a Palaeozoic 'Palaeotethys'; this corresponds to the individualization stage of the alpine geosynclinal domain with its evaporites. During the Jurassic: Atlantic-type spreading continued; this corresponds to the static stage of the alpine geosynclinal period with its radiolarites. During the Cretaceous: compression occurred according to a Pacific model with, on one side (generally Afroarabian), island arcs involving ophiolitic nappes (palaeotectonics), on the other side (generally Eurasian), Andean-type cordilleras; this is the orogenic stage of the Alpine geosynclinal period with its flysch. During Early and Mid-Tertiary: collision of Eurasian and Afroarabian plates and its consequences, with two main episodes: (i) at the end of the Eocene, the main collision leading to Alpine thrust tectonics involving the precedent palaeotectonic structures; (ii) during the Oligocene and Miocene, large basement folding of the continental margins then in contact, deforming both autochthon and allochthon; this is the late geosynclinal period with its molasse troughs. The Late Tertiary (Pliocene and Quaternary): birth of the Mediterranean Sea, announcing a new evolutionary phase, quite different from Alpine history.

Chronologically the three main stages are: Triassic with the beginning of the mesogean ocean spreading; Jurassic–Cretaceous boundary, with the beginning of compressive phenomena, subduction, island arcs and cordilleras (so-called 'Late Jurassic revolution'); Miocene–Pliocene boundary with the extending (mainly) opening of the present Mediterranean Sea (so-called 'Pliocene revolution'). During this evolution there were successively three types of oceans: a Mesogean palaeo-ocean of Atlantic-type in Triassic and Jurassic, and then a Pacific-type ocean in Cretaceous and Eocene and thirdly a Mediterranean Neo-ocean (partly) in the Pliocene and Quaternary. In the same way there were two types of island arcs: Cretaceous to Eocene Pacific-type and Plio–Quaternary Mediterranean type: the first with *root subduction*, the second with *front subduction* (Aubouin, 1975). (*Root subduction:* behind the structures and in the same direction; *front subduction:* in front of the structures and in the opposite direction). Mediterranean arcs are of the same character as Caribbean and Indonesian arcs: they appear in the late history of alpine chains. Western Pacific arcs must be compared with an early stage of the alpine chain (see the following succession: Pacific stage during the Cretaceous and Mediterranean stage during Pliocene and Quaternary; Fig. 5).

The above summarized tectonic sequence is the chronology for the middle Mediterranean area. In relation to the configuration of the Mesogean Palaeo-ocean and the Eurasian and Afroarabian continental margins and their motions, the chronology and relative importance of the various tectonic events varies regionally. Towards the east, the late Cretaceous phase grows and its meaning changes: it becomes the main collision phase, earlier here than towards the west (Ricou et al., 1975). In contrast, towards the west, the same phenomenon can appear as late as the end of Eocene. In the same way, Mediterranean evolution seems to be later (Pliocene) in the east than in the west where the Corsican–Sardinian rotation took place during Oligocene and Miocene times.

The structure also differs regionally. We have seen that, in the east, the main foreland is the Afroarabian plate while the Eurasian continental margin, less deformed but highly magmatized, is thrust over it; this asymmetry recalls the present Pacific asymmetry, the Afroarabian border being of west Pacific type, the Eurasian border of east Pacific type. It is not so clear to the west: there is no andesitic–granodioritic magmatism before Oligo–Miocene times; and in the eastern Alps it is Africa which is thrust over Europe: there is a great change at the level of Dinaric transverse structures, and the Scutari–Pec line or one of the others is a major feature of Alpine European tectonics.

References

Angelier, J. (1973). 'Sur la Néotectonique égéenne: failles anté-tyrrheniennes et post-tyrrheniennes dans l'île de Karpathos (Dodécanèse, Grèce)', *C.R. Soc. Géol. France*, 106–8

Argand, E. (1922). 'La tectonique de l'Asie', *XIIe Cong. Geol. Inter., Bruxelles*, 171–372

Argyriadis, I. (1975). 'Mésogée permienne, chaîne hercynienne et cassure tethysienne', *Bull. Soc. Géol. France* (7), **XVII**, 56–70

Aubouin, J. (1961). 'Propos sur les géosynclinaux', *Bull. Soc. Géol. France* (7), **III**, 629–702

Aubouin, J. (1965). *Geosynclines*, Elsevier, Amsterdam, London, New York, 335 pp.

Aubouin, J. (1973). 'Des tectoniques superposées et de leur signification par rapport aux modèles géophysiques: l'exemple des Dinarides; palaeotectonique, tectonique, tarditectonique, néotectonique', *Bull. Soc. Géol. France* (7), **XV**, 426–60

Aubouin, J. (1975). 'De la position structurale des zones de subduction: subduction frontale et subduction radicale', *C.R. Acad. Sci. (Paris)*, **281 D**, 99–102

Aubouin, J. and Dercourt, J. (1975). 'Les transversales dinariques dérivent-elles de palaeofailles transformantes', *C.R. Acad. Sci. (Paris)*, **281 D**, 347–50

Aubouin, J. and Durand Delga, M. (1971). Méditerranée (Aire), *Encyclopaedia Universalis*, Vol. X, p. 743–6

Aubouin, J. and Guernet, C. (1963). 'Sur une tectonique transversale dans le Pinde méridional au parallèle de Karpenission (Grèce)', *Bull. Soc. Géol. France*, (7), **V**, 77–78

Aubouin, J., Mattauer, M. and Allegre, C. (1976). 'Les nappes ophiolitiques du Sud-Ouest Pacifique et l'interprétation des chaînes alpines', to be published, *4e Réunion Annuelle des Sciences de la Terre, Soc. Géol. France*, Paris

Bergougnan, H. G. (1975). 'Relations entre les édifices pontique et taurique dans le Nord-Est de l'Anatolie', *Bull. Soc. Géol. France*, (6), **XVII**, 1045–57

Bourcart, J. (1962). 'La Méditerranée et la révolution pliocene', *Livre Mémoire Paul FALLOT* 1, 103–16, *Mem. Hors série, Soc. Géol. France*

Caputo, M., Panza, G. F. and Postpichl, D. (1970). 'Deep structures of the Mediterranean basin', *Jl geophys. Res.*, **75**, 4919–23

Dercourt, J. (1970). 'L'expansion océanique actuelle et fossile', *Bull. Soc. Géol. France*, (7), **XII**, p. 261–309

Dercourt, J. (1972). 'The Canadian Cordillera, the Hellenides and the sea floor spreading theory', *Canad. Jl Earth Sci.*, **9**, 709–43

De Roever, W. P. (1957). 'Sind die Alpinotypen Periodtitmassen vielleicht tektonisch Verfrachtete Bruchtücke der Peridotitschale?', *Geol. Rundschau, Bd.*, **46**, 137–46

Dewey, J. F. and Bird, J. M. (1970). 'Mountain belts and the new global tectonics', *Jl geophys. Res.*, **75**, 2625–47

Dewey, J. F., Pitman, W. C., Ryan, W. B. F. and Bonnin, J. (1973). 'Plate tectonics and the evolution of the alpine system', *Bull. Geol. Soc. Amer.*, **84**, 3137–80

Dietz, R. and Holden, J. C. (1966). 'Miogeosynclines in space and time', *Jl Geol.*, **74**, 566–83

Drake, C. L., Ewing, M. and Sutton, G. H. (1959). 'Continental margins and Geosynclines: the East coast of North America north of Cape Hatteras', in Ahrens, L. H., Press, F., Runcorn, K. S. and Urey, H. C. (eds.), *Physics and chemistry of the Earth*, Vol. 3, Pergamon Press, Oxford, 110–98

Durr, S. (1975). 'Uber Randbereich der Kykladen und des Menderes Kristallins', *Ve Colloque égéen;* to be published, *Bull. Soc. Géol. France*, 1976

Enay, R. (1972). 'Palaéobiogéographie des Ammonites du Jurassique terminal (Tithonique/Volgien/Portlandien s.l.) et mobilité continentale', *Geobios*, **5**, 355–407

Fourquin, C. (1975). 'L'Anatolie du Nord-Ouest: marge méridionale du continent européen; histoire Palaéogéographique tectonique et magmatique durant le Secondaire et le Tertiaire', *Bull. Soc. Géol. France*, (6), **XVII**, 1058–70

Godfriaux, I. (1962). 'L'Olympe: une fenêtre tectonique dans les Héllenides internes', *C.R. Acad. Sci. (Paris)*, **255**, 1761

Katsikatsos, G. (1971). 'L'âge du système Métamorphique de l'Eubée Méridionale et sa signification stratigraphique', *Prakt. Acad. Athènes*, **44**, 228–38

Kay, M. (1951). 'North American geosynclines', *Mem. Geol. Soc. Amer.*, **48** (reprinted in 1955 and 1956)

Le Pichon, X. (1968). 'Sea-floor spreading and continental drift', *Jl. geophys. Res.*, **73**, 3661–97

Le Pichon, X., Needham, H. D. and Renard, V. (1973). 'Traits structuraux de la fosse nord égéenne', *1ère Réunion Ann. Sciences de la Terre*, Paris, 266, *Soc. Géol. France*

McKenzie, D. P. (1970). 'Plate tectonics of the Mediterranean region', *Nature*, **226**, 239–43

McKenzie, D. P. (1972). 'Active Tectonics of the Mediterranean region', *Geophys. Jl r. astron. Soc.*, **30**, 109–85

Mercier, J., Bousquet, B., Delibasis, N., Drakopoulos, I., Keraudren, B., Lemeille, F. and Sorel, D. (1972). 'Déformation en compression dans le Quaternaire des rivages ioniens (Céphalonie, Grèce). Données néotectoniques et sismiques', *C.R. Acad. Sci. (Paris)*, **275 D**, 2307–10

Moores, E. M. (1969). 'Petrology and structure of the Vourinos Ophiolitic Complex of Northern Greece', *Bull. geol. Soc. Amer.*, **80**, 3–74

Morgan, W. J. (1968). 'Rises, trenches, great faults and crustal blocks', *Jl. geophys. Res.*, **83**, 1959–82

Ninkovitch, D. and Hays, J. (1972). 'Mediterranean island arcs and the origin of high potash volcanoes', *Earth & planet. Sci. Lett.*, **16**, 331–45

Ricou, L. E. (1972). 'Le croissant ophiolitique periarabe. Une ceinture de nappes mises en place au Crétace supérieur', *Rev. Géograph. phys. Géol. dyn.*, **XIII**, 327–50

Ricou, L. E., Argyriadis, I. and Marcoux, J. (1975). 'L'axe calcaire du Taurus, un alignement de fenêtres arabo-africaines sous des nappes radiolaritiques, ophiolitiques et métamorphiques', *Bull. Soc. Géol. France*, (7), **XVII**, 1024–44

Smith, A. G. (1971). 'Alpine deformation and the oceanic areas of the Tethys, Mediterranean and Atlantic', *Bull. geol. Soc. Amer.*, **82**, 2039–70

SOME MODELS ILLUSTRATING TECTONIC AND OTHER PROCESSES IN THE LITHOSPHERE AND UPPER MANTLE

H. RAMBERG

Institute of Geology, Division of Mineralogy and Petrology, University, Uppsala, Sweden

As an introduction we may look briefly at possible ways of classifying processes in the crust and mantle. According to a purely geological classification we speak about (i) magmatic processes, (ii) metamorphic processes, (iii) metasomatic processes and (iv) tectonic processes. (We are not concerned here with exogenic processes.)

However, we could also use a more fundamental classification, based on the physical nature of processes, and in particular on the nature of the driving forces. In this scheme we have (i) mechanical processes, (ii) chemical processes, (iii) thermal processes, (iv) nuclear processes, (v) electrical processes.

The *mechanical process* in geology is exemplified by flow of magmas, deformation and flow of crystalline rocks, cataclasis of rocks. The driving force is the mechanical potential as manifested in the form of pressure gradients, deviatoric stresses and body forces (gravity).

The *chemical processes* are driven by forces which are distinctly different from those propelling mechanical processes. Chemical potential gradients on micro- and macro-scales are the driving agencies of chemical processes. The potential can be related to concentration gradients, to pressure gradients, to temperature gradients and to the physicochemical nature of the phases involved. In the present connection I consider the chemical process in very general terms. It comprises not only chemical reactions but also melting and other phase changes as well as diffusion of atoms, molecules or ions over shorter or longer distances.

The *thermal transport process* is simply the flow of heat down temperature gradients. Thus, the temperature gradient is the driving 'force' of the thermal process.

Nuclear processes are propelled by little-understood nuclear forces. Nuclear processes are, however, exceedingly important: the release of latent nuclear energy is probably the driving energy of the changing earth, this basic energy being transformed into mechanical potential, chemical potential, thermal potential, for example, thus supporting mechanical, chemical and thermal processes.

When geological processes and phenomena are discussed it is fruitful to have both the geological and the basic physical classification in mind.

A *tectonic process*, for example, is generally a complicated combination of mechanical, chemical and thermal processes. The movement and deformation of rock masses are mechanical in the sense that they are driven by unstable pressure gradients or body forces (gravity). For a rock to be able to yield plastically, however, or even to slide along a thrust plane, chemical processes must take place either within the body of the rock or within the individual mineral grains. Rock deformation, unless it is strictly cataclastic, involves recrystallization, neomineralization and diffusion. Even tectonic brecciation and low-temperature mylonitisation involve a great deal of diffusion and related chemical transport phenomena. The reason for the coupling between the mechanical part of the deformation process and the chemical part is that deviatoric stresses generally create chemical potential differences which in their turn are responsible for the chemical reactions and the chemical transport. Stress is an excellent chemical catalyst.

This leads to the important conclusion that if a rock is too strong to yield by fracturing and movement of the fragments in 'bean-bag' fashion, it can yield to stresses by chemical activation, diffusion and deactivation. Sometimes the latter process is in the form of dissolution in a pore fluid, diffusion through the fluid in the intergranular spaces and reprecipitation.

I believe that the study of rock structures and textures shows that the kind of chemical processes sketched above play a decisive rôle in the deformation history

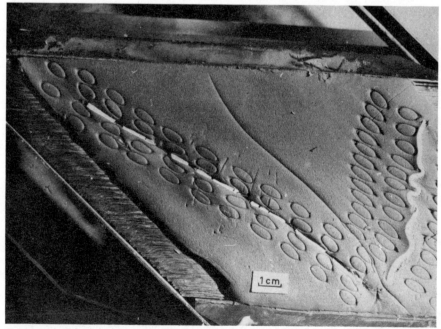

Fig. 1. Boudins and buckle folds formed in two white sheets of 'incompetent' modelling clay embedded in 'incompetent' putty. The system has undergone simple shear during which the upper edge was moved towards the left. Note the rather regular length of the boudins

Fig. 2. Chevron-type folds formed during simple compression of a multilayer of sheets of modelling clay with somewhat different rheological properties. From Ghosh (1968)

Fig. 3. Dome of silicone putty mushrooming up through an overburden of layered modelling clay and painter's putty. Note well-developed rim synclines. Run in centrifuge. From Ramberg (1967)

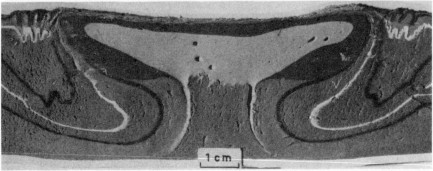

Fig. 4. Example of mushroom-shaped dome of silicone putty in an overburden of modelling clay, etc. Note buckling in front of the lobes of the dome and deep rim synclines. Run in centrifuge

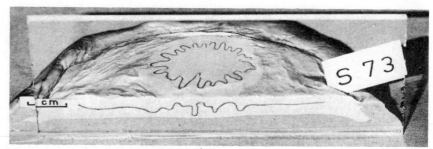

Fig. 5. Vertical and horizontal cut through model dome produced in centrifuge. Note buckling of thin sheet of competent modelling clay in silicone dome. From Ramberg (1967)

of crystalline rocks. Unfortunately, however, we have no real knowledge of these processes. I firmly believe that the lack of knowledge and understanding of the rheological behaviour of rocks in their natural environs is a real bottleneck for the evolution of sound geodynamic models. Much research is needed in this field. But the necessary kind of experimental research is both expensive and difficult.

There is, however, another important aspect of rock deformation which is easier to study experimentally and in which we have seen considerable progress during the last couple of decades. This is the experimental investigation of the evolution of the geometric pattern which heterogeneous rock complexes achieve under stress. Hand in hand with the experimental progress, theoretical work has become increasingly useful to the structural geologist. I am of course thinking of the evolution of folds, boudins, diapirs, orogens etc. by using model materials to imitate rocks and exposing the models to forces which we feel are similar to those encountered within the Earth's crust and mantle. This kind of study throws light on the process of evolution of small- and medium-scale structures as well as global-scale structures such as continental drift and the opening-up of the Atlantic.

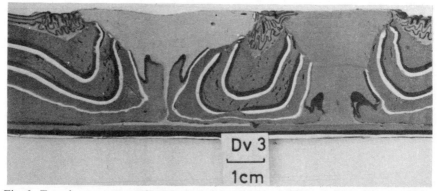

Fig. 6. Two domes, one mushroom-shaped, one block-shaped, of silicone putty in overburden of modelling clay and painter's putty. Note intense buckling of layered overburden adjacent to the domes. Difference in shape of the two domes due to different competency. Centrifuge model made by F. K. Sjöström

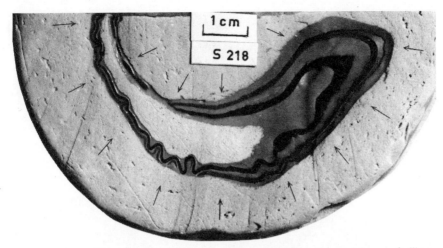

Fig. 7. Horizontal cut through model of anticline formed by buoyant layer of silicone putty containing two competent sheets of modelling clay. Only half of the circular model shown. Note the buckling of the outer part of the embedded competent sheets, and the boudinage of the inner rim of these sheets. The 'basement' in the core of the anticline is exposed in parts of the anticline. Arrows indicate flow of buoyant layer underneath the overburden. For a similar model see Ramberg (1967), p. 118, model S48

No words can describe the results of this kind of experiment as well as the photographs of the models at various stages of development. In conclusion, therefore, a variety of models is illustrated, with brief comments. The sequence begins with small-scale structures such as folds and boudins and progresses through diapirs and batholiths to orogens and sea-floor spreading and continental drift. Figs. 1 and 2 show models of small-scale structures, Figs. 3–7 show dynamic models of medium-scale structures connected with the rise of unstable layers in the field of gravity. In these models many interesting small-scale structures such as boudins and folds are formed during the rise of the domes and the sinking of parts of the overburden. Figs. 8 and 9 show how orogens and orogen-connected

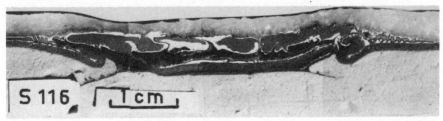

Fig. 8. Model of orogen formed in centrifuge. A triple layer of white, grey and black silicone putty was placed under a sheet of painter's putty with a little higher density to give buoyancy to the silicone layers. Thin sheets were placed on the top to simulate sediments. Different kinds of nappes and folds were produced in the body-force field of the centrifuge. From Ramberg (1967)

Fig. 9. Model of batholiths and granite diapirs in orogens. The structure formed from un-stable horizontal layering of various model materials run in centrifuge. From Ramberg and Sjöström (1973)

Fig. 10. Centrifuged model of continental drift and sea-flow spreading. A relatively stiff plate of wax covered with a dark sheet of modelling clay was fragmented in the stress field caused by drag from two subcrustal diapirs. The diapirs spread below the crustal plate and exerted tension which caused the plate to fracture and move horizontally. Vertical cut shows the two mushroom-shaped diapirs. From Ramberg and Sjöström (1973)

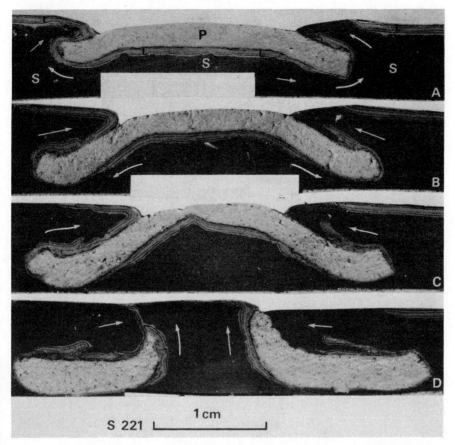

S 221 |————1 cm————|

Fig. 11. Model showing the pattern of subsidence of a relatively heavy plastic plate through a less dense plastic material in a body-force field, here created by a centrifuge. From Ramberg (1967)

batholiths may form due to gravitational instability, while in Fig. 10 we try to reproduce a global-scale phenomenon, *viz* continental drift in a dynamic model. The subsidence of heavy layers in the earth must be an important geodynamic process. Such a process is also modelled in our centrifuge, Fig. 11.

References

Ghosh, S. K. (1968). 'Experiments of buckling of multilayers which permit inter-layer gliding', *Tectonophysics*, **6**, 201–49
Ramberg, H. (1967). *Gravity, Deformation and the Earth's Crust*, Academic Press, New York, 214 pp.
Ramberg, H. (1971). 'Dynamic models simulating rift valleys and continental drift', *Lithos*, **4**, 259–76
Ramberg, H. and Sjöström, H. (1973). 'Experimental geodynamical models relating to continental drift and orogenesis', *Tectonophysics*, **19**, 105–32

Quaternary Europa

PROBLEMS AND AIMS IN QUATERNARY EUROPE

G. F. MITCHELL

Trinity College, Dublin, Republic of Ireland

The average geologist looks essentially backwards, as he tries to unravel the history of the Earth, but the Quaternary scientist must look forward to the developing environment, as well as backwards to the very recent past. In ancient Rome we would have had no difficulty in deciding to which member of the pantheon of gods we should address our appeals for enlightenment—they could go only to Janus, the two-faced god, one face looking forwards, and the other turning back. We would have set up a cult of our own, that of *Janus Quaternarius*, with one face looking back to the geology of the Pleistocene, and the other looking forward to the environment of the Holocene. We could represent him as holding a sword, with the date 10,000 B.P. emblazoned on it, symbolic of the dividing line between the two periods.

Thus, even in the design of our emblem itself, we would emphasize our differences from other geologists, especially in the time scale with which we are concerned, for we speak in terms of at most a few Ma, rather than hundreds or thousands of Ma. Our time scale is short, and evolutionary processes move only slowly, and thus we cannot hope to find in our brief record the helpful appearance of new organisms that so greatly aids our hard-rock colleagues in the positioning of their golden spikes. We must measure in physical units, rather than evolutionary ones. We do not, of course, ignore the fossil evidence. Our era appears to consist of alternating warm and cold stages, the cold stages being rather longer than the warm ones. If we can plot out the record of climatic change, we will have gone a long way to unravelling our problem. In such plotting out, the tracing of a succession of warm and cold floras and faunas will be most important. Unfortunately the evidence on which we base floral and faunal developments rests largely on the remains of flowering plants, vertebrates, beetles and molluscs, all relatively mobile organisms, and while we can often see that there was movement of these organisms provoked by climatic change, it is very difficult to arrange these changes in their correct chronological sequence.

Warm conditions seem to have marked most of the Earth's history. Cold is the badge of the Pleistocene. Where are we to drive in the golden spike that will mark the opening of the Pleistocene? About 25 Ma ago temperatures all over the Earth started to fall in an irregular way, periods of falling temperature being followed by periods of recovery, but always with an overall downward trend. By about 13 Ma ago cooling had reached the point at which icecaps could begin to form in polar regions; though the amount of ice at the poles has fluctuated since then, it seems that at all times there was some ice present. A recent book speaks of *Late Cenozoic Glacial Ages*.

It is useful to remember that this part of the geological record was first studied in northwest Europe, and that it was inevitable that thinking in this area should have greatly influenced concepts of the manner in which it was appropriate to divide the Pleistocene from the Pliocene. When remains of arctic plants were first found in temperate Europe, the existing latitudinal contrast between the forests of the south and the tundras of the north immediately came to mind. A vertical contrast in the geological column between a biozone with woodland fossils below, and one with tundra fossils above, immediately presented itself as a potential horizon for the golden spike.

So the boundary could be placed at the level where the final phase of the archaic—to our eyes—Pliocene woodlands gave way to the first spreading of the Pleistocene tundra, a change which has been shown in the Netherlands to have taken place about 2 Ma ago. This date lies not too far from the start of the Gilsa palaeomagnetic event, which appears to fall about 1,800,000 years ago. This is the same order of age as the base of the Calabrian, when cold-loving organisms first made their appearance in the Mediterranean Sea, and of the Apsheronian in the USSR when *Archidiskodon meridionalis* appeared in Georgia and Azerbaijan. It can also be linked up with the rather earlier events in the Olduvai Gorge, and so give us important contact with early man. If, after further stratigraphical testing, and further refinement and standardization of sampling and processing, palaeomagnetic methods are shown to have sufficient validity for our purposes, we would do well to define our divisions in palaeomagnetic terms rather than pursue the will-of-the-wisp of type sections. Type sections within regions: Yes; global type sections: No.

Many of the older cold stages of the Pleistocene seem to have been periods of frost rather than ice, and the great ice-masses seem to have come later. The beginning of the Brunhes epoch about 700,000 years ago should give us another valuable horizon, and the first appearance of ice across northern Europe from England to the Soviet Union may well lie at something of this order of time. It has been suggested that a palaeomagnetic event can be demonstrated at about 10,000 B.P., and if this can be shown to be the case, then palaeomagnetism and radiocarbon dating could go hand-in-hand in defining the Pleistocene/Holocene boundary.

It seems essential to use a physical dating for this boundary. In our myopic European vision it lies where Pleistocene tundra gives way to the uninterrupted return of warmth-loving woodlands, and we think that some analogous change should be recorded throughout the world at this level. We talk of Pluvial Periods, but we find it hard to decide whether much of Africa was getting wetter or drier at this time, or indeed remaining unchanged. What change can we hope to trace in the equatorial rain-forests, or in those parts of the Earth that are still ice-bound?

What of our present position? I think that few of us feel that the cold of the Pleistocene has gone forever, and therefore we take ourselves to be living in a warm stage, or 'interglacial'. We recognize that about 6000 years ago conditions seem to have been warmer than they are today, and we speak of a 'Climatic optimum' or 'Hypsithermal interval'. Climate seems to have been favourable again about A.D. 1300, and then there was deterioration into a 'Little Ice Age'. There was a further rallying in the beginning of the current century, but recently there have been signs of deterioration, and gloomy prophets have been calling 'Repent ye, for the Kingdom of Ice is at hand'. But even one hot summer has been sufficient to make the Doomsday merchants run to hedge their bets.

It is curious the way fashion swings in our subject, just as it does in all others. The first concept of the Pleistocene was one of warmth, interrupted by four short episodes of cold. Penck considered that the Mindel/Riss interglacial lasted for almost 250,000 years, and so it became known as the Great Interglacial; following on the work of Milankovitch, Zeuner considered that figure too great, and he reduced it to 190,000 years. Now the pendulum is going the other way, and we envisage the Pleistocene as cold, interrupted by short phases of warmth. As many as eleven periods of cold are now reckoned on, and the warm stages have got shorter. In recent years, 30,000 years has been envisaged as the duration of the last warm stage (Ipswichian), while today ideas are beginning to come forward that no more than 10,000 years is the typical length of a warm stage. If this is correct, we have had our 10,000 years, the prophets of doom are right, and we had better all order our fur coats.

But suppose we are not living in an ephemeral warm stage, suppose the cold of the Quaternary Era is over and done with, what then? As I have already said, warmth rather than cold seems to be characteristic of the Earth. Are we entering the Quinary Era, when the glorious warmth that marked the Tertiary will return, and dense and monotonous woodlands will endeavour to re-establish themselves over higher latitudes in both hemispheres?

One astronomer at least believes that this is the case. Most galaxies have a pair of visible spiral arms, and along the edges of these arms there are dark lanes of dust and gas. When our Solar System enters such a dust lane, the sun accretes dust particles, its surface layers rise in temperature, and radiation output rises. If we believe that increased cloud cover lowers the Earth's surface temperatures, then the increase in radiation will increase the albedo of the Earth, and cause loss of heat greater than that brought in by the increased radiation, and a cold stage will follow.

It takes our Solar System 500 Ma to make a complete orbit round the Milky Way galaxy, and during that time it twice passes through a spiral arm. Professor McCrea points out that the pre-Quaternary ice ages appear to be separated from one another by about 250 Ma; each of them lasted for several Ma, and during them conditions varied from warm to cold, just as has happened more recently. 250 Ma is the time it takes the Solar System to travel from one arm of the galaxy to the other, and several Ma is the time it takes the system to pass through an arm.

Today the Solar System is on the edge of a spiral arm, having emerged from the associated lane of dust and gas about 10,000 years ago, just the time that we non-astronomers have chosen to bring the Pleistocene cold stage to an end. Thus according to McCrea we are out of the arm, and the climate is set fair for a warm Quinary Era, which will last, unless palaeontologists are allowed to chop it up into smaller pieces, for at least 250 Ma.

But before we float off into Quinary warmth, let us look back again to the Quaternary deposits, which still hold vast amounts of information, as yet unrevealed to us. We are very fortunate that our deposits are still relatively intact, unblurred by the passage of time and not yet deformed by metamorphic processes. This privileged position makes it imperative for us to study them in the greatest detail possible, using all forms of experimental techniques, because it is not impossible that new techniques, first applied to young and undeformed material, may also prove capable of application to older strata, and may coax them to disgorge information hitherto thought to have been lost beyond recall.

For many years our studies were confined to deposits which were easily accessible on the surface of the Earth, or could be reached by drills firmly located on solid materials. Today the oceans and the lakes are beginning to reveal their secrets.

It is now possible to recover cores more than 20 m long from ocean floors at depths of more than 3000 m, and to study the oxygen isotope ratios and the magnetic stratigraphy of the sediments. Our Japanese colleagues have recovered a core 200 m in length from a lake where the water depth exceeded 60 m, and the palaeontology, sedimentology, limnology and chemistry of the sediments are all being closely studied. The presence of layers of volcanic ash in the core means that fission-track ages, as well as palaeomagnetic ages, can be determined.

When we have found and studied the deposits, we must describe our results with precision, using terms whose meaning is internationally recognized. This is one of the areas where we are weakest, though the fault perhaps does not lie entirely with us, but rather with the processes of natural selection which have made us the individuals we are. Why do we become scientific research workers rather than factory workers on an assembly line? It is because we have been blessed, or cursed, with a natural curiosity, to seek out new things, or find a new way of doing old things. Each of us is convinced that our way of doing or describing things is better than that of anyone else, and we resist all attempts to make our actions or our words conform to those of others. This may be great fun, but it makes for very bad science.

Consider another part of the modern world, that of administration. Just at present the countries that many of us come from are trying to weld themselves into a European Economic Community. To achieve this many internationally binding regulations will be necessary, and it is essential that these are spelled out in words whose particular meaning will survive translation into several different languages, and can be ruled on if necessary in the courts of several countries. We would be well advised to adopt some similar uniformity of vocabulary. The international Quaternary body, INQUA, has a poor record in this respect. In 1953, in Rome, a commission was established to prepare a Dictionary of Quaternary Terminology, but it was dismissed eight years later in Warsaw, because no progress was being made. Since then there has been little activity in this field. A lexicon is badly needed.

Also we want clear definitions of our boundaries, with stratotypes for each region, and not the continued use of vague terms like Würm and Weichsel. We can excuse Penck for not anchoring the word Würm to a type-site, because in his day such matters did not carry the same importance, but there is no excuse for failing to attach basic type-sites to the north European terms, Elster, Saale and Weichsel, and if this cannot be done, we must abandon these names. This is a matter to which the INQUA Stratigraphic Commission is giving attention. We want detailed descriptions of all published sites, stratotypes and others, drawn up in accordance with the stratigraphic code of the IUGS Subcommission on Stratigraphic Nomenclature.

When these matters have been set in order we can hope to make some progress in correlation. This is a matter that has concerned INQUA since its earliest days, and as far back as 1932 the Leningrad Congress drew attention to the importance of correlating the Quaternary deposits of Europe. But, as I have said before, we are a preselected group of individualists, and we each know better than anyone else. Before the INQUA Congress in Madrid in 1957, Dr. van der Vlerk sent a request for a stratigraphical table to 22 different countries. The 22 tables received proved to be completely incompatible with one another, and although in Leningrad it had been agreed to divide the Quaternary into four main units, most countries did not employ the units, and where they were used, each specialist interpreted them in a different way. In general, the position is not much better today. However, the Geological Society of London has recently published a scheme of correlation for the Quaternary deposits of the British Isles, based on type-sites around which local schemes with local terminologies have been set up. It is to be hoped that other countries will soon follow suit.

I have already referred to the important part that palaeomagnetic studies will play in this work. Radiometric dating will also be important. Radioactive carbon carries us back about 40,000 years; with coaxing and further study thorium and proactinium will cover the last 250,000 years, but then there is an awkward gap—which we must try to fill—until we get back to the millions of years ago when potassium and argon come into play. If we want to forecast the future, we must have still more refined datings from the very recent past, and we must seek to eliminate the discrepancies between radiocarbon dating and dendrochronology, or at least arrive at a standard method of correction to harmonize the two.

The oxygen isotope method will probably become more and more important. In deep-sea records we can now follow the changes in foraminiferal tests below the Brunhes–Matuyama boundary 700,000 years ago. Soon the records from Greenland ice should go back to 100,000 years. It is just beginning to be realized that the same precipitation that feeds the ice of the ice-masses also feeds the wood of the trees, and that wood should have the same record of climatic change built into it. Dated bristle-cone pine-tree rings now stretch back over 8000 years, and if refined techniques can also reveal the oxygen-isotope record of the rings, we have not only a source of meteorological information, but also the possibility of linking the record of the lands with that of the oceans.

More detailed investigations of wood are revealing that not only is there the relatively obvious change in the breadth of the rings from year to year, but that the density of the wood formed also varies as the season progresses. Not all varieties of tree are suitable for the counting of rings, which is chiefly successful in climatic regions with a marked seasonal rainfall, and detailed studies of wood densities may enable the geographical range of wood studies to be widened enormously.

The study of organic remains will be carried into greater and greater detail. Palynology has come a long way since von Post first started his work about sixty years ago. He was content to take his samples relatively far apart, and to count in each sample about 100 pollen grains derived from perhaps six varieties of trees. Now pollen diagrams are being published with samples at 1 cm intervals, 500 or more pollen grains counted in each sample, and as many as 40 taxa shown in the diagram. In this very greatly widened range of identifications the stereoscan microscope has, of course, played an important part, but a considerable amount had been done before it came into general use.

The stereoscan microscope has brought about a revolution. The identification of many pollen grains can now be brought to the level of the species rather than remaining at that of the genus, and the identification of other small fossils, diatom frustules for example, has been similarly improved. Macrofossils too have benefited; the ornament on seed-coats can now be seen in much greater detail, and many problems of subspecies and varieties are beginning to be cleared up.

More detailed scrutiny generally is widening the range of other plant parts, such as wood and cuticle, which can now be closely identified, and a better picture can be built up of the environment from which the fossil material is derived. Most fossil material has decayed to some extent, and has often been extensively invaded by fungi, which in many cases have left a large range of debris behind them. Some Dutch workers are now starting to pay especial attention to these fungal parts, and new information is beginning to emerge.

But among the wealth of vegetable parts that can be studied, the pollen grains will probably continue to hold their leading position. When von Post first worked with them, they were principally thought of as offering special possibilities for the dating of the deposits in which they occur. But when later, thanks to the work of Iversen, it came to be realized how profoundly man from Neolithic times onwards had been capable of altering the vegetation cover, and therefore the pollen grain content, the value of pollen studies for dating purposes slumped rapidly. But as this emphasis lessened, their importance for reconstructing plant assemblages of the past, and therefore past environments, grew rapidly. For more general purposes their chief importance will be for indicating overall climatic swings from cold to warm, and back to cold again. The amount of pollen produced per unit area will rise and fall with the climate, and thus the quantity of pollen in a sediment will be just as important as its nature. Every effort must be made to enable meaningful absolute pollen counts to be made in as wide a variety of sediments as possible.

This involves further study of pollen production, pollen transport, pollen sedimentation and pollen preservation.

The animal kingdom must give its yield of information also. Mammalian faunas have long been studied, but here the rarity of preservation must mean that information will always be irregular. Still, from time to time, we do get lucky bonuses, as for example the recent discovery of a rock-fissure at Westbury-under-Mendip, which produced remains of at least 30 different mammals, apparently of Cromerian age.

Under the leadership of Professor Shotton in Birmingham, the study of Pleistocene coleoptera has taken on a new lease of life, and it is now clear that detailed studies of beetle faunas can yield an immense amount of climatic information. Here too the use of the stereoscan microscope can widen the range of identification, and make those identifications more certain. There must be other invertebrate groups still lurking in the wings, and awaiting the *impresario* who will coax them on to the stage to reveal new and important information.

If all these technical advances, at which we have taken a rapid glance, were to be developed to their logical conclusions—to say nothing of other still more sophisticated techniques as yet unborn—it is obvious that we would have a vast amount of information on hand, so vast that isolated human brains could not handle it, and elaborate statistical computerized programmes will have to be devised. This is, of course, a field in which COGEODATA is very active.

But supposing we have all this information, and supposing we have unravelled the complexities of the history of the Pleistocene, to know itself is not enough. Our knowledge must be put to work. One hundred years ago knowledge for knowledge's sake was still enough, but this is a position today's hungry world can no longer afford. For too long too many scientists, geologists included, have pursued an esoteric investigation, reached a result, communicated the result to their colleagues in a cosy club-like atmosphere, and forgotten all about that investigation, as they happily proceeded to another. We should bear in mind that we must be more than prosperous members of an exclusive club, the club of developed countries, the Club of Rome. We must see ourselves as the foundation of a new type of community, the European Economic Community, which we hope will struggle to make the most co-operative use of its natural resources, and in doing so will both set an example, and be prepared to give every practical assistance, to other less favoured, less developed parts of the world.

UNESCO has undertaken several major international scientific projects, and its geological programme, the International Geological Correlation Programme, is just beginning to gather pace. Two periods of the Earth's history have been singled out for special study, the Precambrian and the Quaternary. The Precambrian was chosen because of its mineral wealth, and the Quaternary because it holds the present environment and its most recent past. And this is not a programme primarily to gain knowledge for knowledge's sake; as far as the Quaternary is concerned it is to study the past in order to assist the present and to predict the future.

Current processes on the Earth's surface are being monitored in greater and greater detail, as the names of the UNESCO programmes indicate: The International Biological Programme, the International Hydrological Decade, Man and the Biosphere. Detailed studies of the atmosphere and its climate are being pursued on a world-wide scale, and past climatic records are being every more closely scrutinized to form a bridge with the future. As I have already noted, information is pouring in from all sides, and these floods of knowledge must be harnessed, so that they can best enter into the service of mankind. New multi-variate statistical techniques will have to be wedded to the powers of new high-speed, high-capacity computers, so that the most profitable analyses and syntheses can be arrived at.

Let us look at some of the Quaternary areas in which the Board of the International Geological Correlation Programme has called for action.

Deposits. Glacial, periglacial, freshwater and marine deposits must all come under review. Terrestrial Quaternary deposits are usually loose and unindurated, and so offer problems in urban planning and in engineering construction. On the other hand they themselves often provide large quantities of raw materials for building purposes. Many of them are relatively coarse in grain, and hold important economic supplies of groundwater. In many parts of the world, for example North Africa and Australia, there are large deposits of fossil groundwater, which have to be regarded as non-replaceable economic deposits, and must be mined with the same care as a valuable metallic ore. The ever-changing alluvial deposits of the Bengal delta carry one of the most dense populations of mankind, a population at the mercy of some of the world's largest rivers and some of the world's most powerful storms. Approaches to ameliorate the problems there must be many-faceted; flood-control, increase of rice production and maintenance of supplies of fish must be carefully balanced against one another, if one cure is not to produce another disease.

Landforms. Most of the landforms we see around us today received their final shaping during the Quaternary. Landforms can often effect a remarkable level of control on land suitability for rural and urban planning, and costly mistakes have been made where this factor has not been recognized. We must see that the less-developed countries profit from our errors. Correlation, description and cartographic representation of landforms are of particular importance for present and future human activity.

Soils. The soils of the Earth, which are its most universal and most important economic deposit, are mainly developed on Quaternary materials. Unlike fuels and minerals, they need not be exhausted by exploitation, and their proper management is vital for mankind. We still do not know enough about the way the parent materials become transmuted into soil, and there should be major investigations into genesis and distribution, not only of recent soils, but also of palaeosols. We now realize that many of the richer soil areas of the Earth have been occupied by farmers for periods often much longer than 5000 years, and that these human operations have profoundly changed the primary nature of the soil, and have moulded its later development. Detailed studies of interglacial palaeosols which were not disturbed by human activity, and enjoyed a much longer period of natural evolution, should throw a great deal of light on more recent soil development. Many low-grade ores are often only profitable to work in those zones which have been secondarily enriched; detailed study of Tertiary and interglacial weathering processes might help us to understand the mechanisms that brought about enrichment of older metallic deposits.

Water. Petrol today costs about 75 pence per gallon, or rather less than 1 Deutsche mark per litre. I always tell my students that water is a much more remarkable fluid than petrol, and that if it was only available at the same price as petrol, we would all appreciate it more, and make much more efficient use of it. At such a price industry would recycle its water, and would no longer discard it like a once-used piece of toilet-paper.

Because it appears to be cheap, we all use it much too freely, often with disastrous results. Take the farmer with his irrigation scheme; he has paid for the water, hasn't he? and he wants to see it sloshing through his fields, so that he gets value for his money. He may get value for his money but he also gets ruin for his soil. The water-table rises, the alkalies rise, and the soil becomes flooded and poisoned. There is no more terrible sight than to fly over parts of India and see the flowing irrigation canals, which, although they did bring an initial prosperity, are now bringing disaster to thousands and thousands of hectares. And this does not happen only in the less-developed countries; today in the Murray River basin in Australia the hydrological engineers are desperately trying to juggle with masses of highly salinated water, trying to retain it in isolating-ponds when river levels are low, and hoping to release it when the flood waters reach a volume which may be able to carry it off diluted below the level at which it can do serious damage. Human health can suffer also; we are only just beginning to realize how important trace elements are in our make-up. In an Indian area recently the rising water-table carried molybdenum up to the surface, where it was absorbed by the staple crop of the area, sorghum. The consequent higher dietary intake of molybdenum brought about a displacement of copper from the human body, and many young adult males became afflicted by crippling weakening of the bony structure of their legs.

We still know all too little about the role in nature of this remarkable and all-important fluid. All-important for all of us, because by weight it makes up 60 per cent of our bodies. Three litres a day are sufficient to keep the human body topped up, like a car battery, but most of the inhabitants of developed countries find it necessary to have a passing contact with 135 litres every 24 hours.

Climate. Water appears again as another of the climatic variables, whether as ice in an icecap, water in a pluvial lake, or vapour in a tornado. Careful analysis of old meteorological records shows the variations of the recent past; detailed study of today's trends may reveal the variations to be expected in the near future. If our aim is prediction, this is one of the most promising areas of research, and a great many people are active in this field.

Sea level. Water again is an important agent here, its volume being inversely proportional to the size of the Earth's ice-masses. During the later part of the Quaternary sea level appears to have fluctuated through a range of at least 150 m, from 125 m below its present level to 25 m above. Were all the ice on Earth to melt, sea level would probably rise by about 50 m, with catastrophic results for large areas of low-lying densely populated lands, where even small changes in level have important consequences for urban planning, harbours, etc. By far the greatest amount of the world's ice lies in Antarctica, and there are some workers who consider that the ice there surges from time to time, suddenly discharging great masses of ice into the ocean, and giving rise to disastrous tsumamis and fluctuations in level; we should perhaps give some attention to this possibility.

Tectonics. Most workers view the apparent Quaternary changes in sea level as due on the one hand to eustatic effects consequent on the growth and decay of ice-masses, and on the other to the isostatic effects of the loading and unloading of the weight of ice on the Earth's crust. Tropical oceanic islands have been used as bases free from isostatic effects to investigate eustatic changes in sea level. But our isotopic colleagues now claim that they can monitor the growth and decline of the ice-masses by changes in ocean temperatures, and say that some of us attribute sea-level changes to eustasy at a time when there is no evidence for a change in the size of the Earth's ice masses. They would say that parts of the geoid do change in shape, rising here and falling there, and that apparent eustatic changes may be due to tectonic deformation of the geoid. There is good evidence that the southern part of the North Sea basin is actively subsiding today; the numerous basins around the British Isles that are now being revealed by exploration for oil and gas show that Tertiary subsidence certainly took place. Seasonal changes want to be looked at very closely, because ability to predict changes would be of the highest importance.

You will have noticed that I have used the verb 'predict' several times, and this is what our aim must be, 'Prediction'. Let us study the past in as great detail as we can, and then, armed with that knowledge, deliberately turn around and try to foresee the future. This is the real aim of the IGCP: the report of the first session of the Board says with regard to the Quaternary 'Accurate correlations within this latest phase of Earth history are a prerequisite for the use of geology as a predicting science with regard to the geological environment of Man'. This is the challenge with which the IGCP Board confronts us; we should do our best to meet it. (The progress of IGCP can be followed in *Geological Correlation*, published by UNESCO, Place de Fontenoy, 75700 Paris, France.)

Yet it will be of little profit for us to know what the future may bring, if the future can never arrive because we have destroyed the present. As Quaternary scientists we *know*—and that knowledge prevents us from putting our heads in the sand and pretending that we don't know—just how fragile the Earth's geological environment is. Every year there are more people on the Earth; every year the techniques of food production get more and more sophisticated: and more and more vulnerable to natural disaster, to climatic variability and to fuel crises. Accurate forecasting may or may not evade us, but there is no point in spending thousands of man-hours and millions of dollars trying to predict what nature will do to our environment, if in the meantime we have ruined it ourselves. As I say we *know* that fragility, and we must force our knowledge on politicians and others.

Some or our colleagues have already made a start on this task, and a group which met in Uppsala early in 1975 drew up a resolution on the importance of environmental Quaternary research as a basis for resource planning.

The resolution runs as follows:

The Uppsala Resolution

Geological Environmental Quaternary Research—the Basis for Resource Planning

(1) The geological environment is the basis of all biological activity, and is also man's home. It will also be his home in the future.

(2) The environment must therefore be treated in a responsible manner, a manner that recognizes long-term planning for man's future.

(3) In all use of the environment by society there must be awareness of the three-dimensional pattern of the Earth's crust and of the working of geological processes, in order that disastrous mistakes can be avoided.

(4) Governments must be made aware that most geological environmental resources are non-renewable, and once consumed cannot be restored. Careful planning and documentation must precede exploitation.

(5) In addition to the current studies of biotic environmental problems, of the pollution of surface-water and of the atmosphere, special attention must be paid to disturbances of the geological environment and its natural equilibrium.

(6) Palaeoclimatology, sea-level changes, and evolution of the land surface are related to the geological processes of the environment. Knowledge of their interaction and of present-day processes are necessary to avoid disastrous situations in the future.

(7) In the light of these facts, it is essential that:

(a) in the training of geologists, stress is laid on the problem of environmental planning;

(b) geologists must take part in environmental planning;

(c) geologists must impress on political and technical authorities the fact that if there is to be proper management of the environment the fullest use must be made of geological knowledge;

(d) geologists must use all forms of communication, including the mass media, to impress on the public in general the very important contribution that geological knowledge can make to the solving of environmental problems.

More recently, in Brazil, there was an International Symposium on the Quaternary, attended by almost 100 people from 20 different countries. A resolution very similar to the Uppsala Resolution was passed, with the addition of a further clause which drew attention to the especially fragile nature of tropical and subtropical environments.

As Quaternary scientists we have all enjoyed the exhilaration of field-work, the pleasure of quiet concentration in the laboratory, and the satisfaction of communicating our work to colleagues in an academic meeting. Most of us would prefer to continue to shelter in our ivory towers, venturing forth only occasionally to club-like scientific meetings. But we cannot remain in hiding, while the environment is being wrecked around us.

The question of professional associations for geologists, to give public standing to those who follow the vocation of geologist, has been very much under discussion in recent years. Quaternary scientists should unite, not to gain selfish professional standing, but to provide themselves with corporate strength and a corporate voice with which to speak out and ensure the protection of the environment. The Earth is our home, and we hope it will be the home of our children, and of our children's children.

The Future of European Geology

ORGANIZATION OF EUROPEAN GEOLOGY: PRESENT AND FUTURE

J. M. HARRISON

UNESCO, 7 Place de Fontenoy, 75700 Paris, France

Europe presents in a relatively small compass many of the problems of the world of science and technology, so if Europe can move ahead on the solution of co-operation in geological sciences, all the world will benefit. But before discussing possible organization, we should consider what we wish to accomplish with our capabilities in the field of geosciences. In this paper I shall use the term 'geosciences' instead of geology, for too many people feel that geology is a limited member in the family of earth sciences. I mean the term to include studies on the whole of the geoid on which we live, and for which geology is the integrating force.

As with any science, one of the principal objectives has been, and should continue to be, the advancement of knowledge, in this case about the Earth. I am a great believer in the utility of knowledge, however, and I suggest the second major objective should be to use the knowledge to develop the Earth's resources. For a long time it was enough that a man be recognized as an important scientist for him to obtain a position, probably a professorship at a university, prestige, and research grants. Now, in many parts of world the pendulum has swung, possibly too far in some countries, and the people who supply the funds now want to know specifically what they will get for their money. This attitude is becoming stronger in more countries in Europe and is, I am glad to say, having a most salutary effect on one of the more distressing effects of the high-prestige system.

To come to Europe from Canada as a geologist a generation ago was a bewildering experience. Although we had 'schools of thought' in Canada they were not so firmly founded. One could disagree as a student and not be damned by the 'Professor'. Perhaps I was among the fortunate few but my principal professor when I was an undergraduate encouraged us to disagree with the leading geologists of the day. He had his own theories of course and we were expected to tear them apart too.

To come here and learn of the dominance of so many professors, the strictness with which they held to their own concepts and the way in which they exerted their territorial imperative was a shock. Too much of this still remains. Let me quote from a note written this spring by a younger geologist from Canada who spent part of his sabbatical leave visiting type localities in Europe. 'I was struck by the signifance of "claim rights" held by geologists in specific areas.' Again 'I was disappointed that "certain type localities" had not been studied to death by all geological disciplines. The guardians of such world type localities have an obligation to be sure that everything geological is obtained from them.'

The same geologist also commented that European geologists tend to be inhibited by the psychological block of an international border, a fact which many people have noticed, especially Europeans. It is a particular problem that the small size of so many countries circumscribes the geological studies. Of course there are large territories, such as European USSR, France, the Germanies, but compared with the scale of geological processes, even they are relatively small. It is easy to understand, therefore, why the International Geological Congress developed 100 years ago in Europe (even though an officer of the Geological Survey of Canada was the organizing Secretary General). Out of this organization has come truly remarkable European geological co-operation, especially in the synthesis of different kinds of geological data on geological maps. The effect of European geologists on the rest of the world was enormous, for those were the days of empire. Many smaller European countries were providing the cadres of specialists for studying geological problems in other parts of the world. As a result, the synthesis became rapidly world-wide, but was until recently done almost exclusively by Europeans and North Americans.

Today the situation has changed. In many European countries a major difficulty is going to be, if it is not already, how to justify the support of national schools of geological sciences when the national demand for geological information is so small. Governments are among the more pragmatic of institutions and are unlikely to support for long university activities that are not obviously useful to the country. As long as a country was providing economic geologists for its colonies it could afford to support curiosity-orientated research as well, but with the abrupt ending of the colonial system, governments are looking much harder at the value of courses given at universities.

It seems obvious to me that there will inevitably be a concentration of schools of geology in Europe for few, if any, universities today can be expert in all aspects, and indeed few countries can be expert in all. Here the logic of the situation becomes tangled with nationalism and the question of language, for unless geologists of Europe can work in at least two languages the idea of specialized centres being usefully available to the international community is not likely to be more than a dream. Moreover for a national university to have an international focus its staff and its students must be able to work in more than one language.

There is an important consequence that follows from that line of reasoning, namely the kind of specialization that European schools can expect their governments to support. I find it difficult, for example, to imagine that a small flat coastal country will be able to develop a school in, say, mineral deposits, or petrology of igneous rocks, because these aspects relate very little, if at all, to the national geological environment. And thereby, one obvious problem of Europe is that although geology is the science of the Earth, its teaching and practice tends to depend on parochial considerations.

Having argued that European universities (and nations) will have to specialize selectively, one problem is to determine what they should specialize in. One point is of special importance and it is reinforced every day at UNESCO: with the abandonment of colonialism, traditional geological mapping and geology of mineral deposits has become less important to European geologists. For a time, European schools can be expected to train geologists from developing countries, but for a limited time only. It is essential that schools be established in those countries although, perhaps, on a regional basis. The principles of geology—of all science—are world-wide, but the utilization of the knowledge, as with the teaching, depends on local conditions. Secondly, of what good is it for a student from a small African state to be trained on modern expensive instruments and to spend his efforts on such problems as the mid-ocean ridges, or the petrology of alkaline rocks, when he will be trying to put his knowledge to use in a country that has neither the money nor the technicians for complex instruments, and where he will be faced with problems of his own country's well-being? As one of my Egyptian friends has said, the greatest brain drain is not the loss of scientists to other countries, but the loss to developing countries that results from training scientists to do the science of the industrialized world. Another friend from Saudi Arabia told me recently that Saudis can go abroad for graduate training only if the host university will accept their Saudi problem for the dissertation. This includes geoscience students and it is a procedure that has much more merit than it has drawbacks.

Some of you here will know about the International Centre for Theoretical Physics at Trieste. The ICTP is funded jointly by the government of Italy, the International Atomic Energy Agency and UNESCO, but receives substantial financing for special purposes from other agencies. Professor Abdus Salam of Pakistan, who is also Professor at Imperial College in London, directs the Centre. It caters mainly for physicists from developing countries by giving them fellowships that permit them to visit the Centre for specialized courses of short duration, to participate in study groups, and to carry out their own research. This is relatively easy in theoretical physics for a library, office space, meeting rooms and a computer are about all the equipment needed. Still, it strikes me that the principle could be applied to some aspects, at least, of geology. Scientists in isolation soon become too localized to be fully effective, especially if they are teaching. If they know, however, that every couple of years they could spend, say, three months at a university that specialized in a topic or topics of concern to the work they were doing in their home country I imagine the results would be most useful. First, knowing that they would be able to follow up particular points with their colleagues and that they would have access to sophisticated instruments, their interests would remain broader than the daily problems that beset them. Secondly, they would return home with new impetus, new vigour, and new points of view with which to stimulate their students and countrymen. Thirdly, the feeling of isolation would at least be reduced.

Another possibility stems from an unusual, perhaps unique, institute in Nairobi—the International Centre for Insect Physiology and Ecology (ICIPE).

Why not a regional centre for geological sciences, or some aspects of them, in appropriate locations in Africa, Asia, the Middle East, Latin America, the Caribbean etc? I know there is a first-class school of African geological studies in Leeds—perhaps it could be the basis for building a similar institute in Africa. The ICIPE, which is being watched with interest around the world, is headed by a Kenyan, Dr. Tom Odhiambro, and in four or five years since its creation has built up a good physical plant at the University in Nairobi and is now taking students from other African countries. It is true that most of the teaching staff are expatriates but that will change. It would seem to me that if it can be done for insects it can be done for rocks!

You will have gathered that I believe one of the most important specialities for European schools of geology should be that of helping the less developed countries to help themselves. The UNDP and especially the World Bank, are preoccupied with economic development. Without the trained people—scientists, engineers, technicians—any talk of economic development is arrogant nonsense. One of the most important investments that could be made is in technical education for development, where the UNDP helps substantially but the World Bank not at all. Perhaps geoscientists can do something for earth sciences by using existing structures and institutions. It still will not absolve us from the need to help directly with our own trained staff until the rest of the world can stand on its own. A disinterested federation of European societies could certainly have much influence on governments and aid agencies.

But what of disciplines for European problems? I would like to emphasize some particular aspects that should be of great concern to the fraternity of earth scientists in Europe, and others where European earth scientists should lead the world, as indeed they probably have a responsibility to do.

Europe is more thoroughly mapped geologically than any similar region in the world. If there is any place where geology and geophysics should be combined it is in Europe. Surely geophysicists would benefit from collaboration with geologists on problems of earth science, and I know geologists would learn a great deal if they better understood geophysical techniques. Years ago, when reading the proceedings of geological meetings that took place in North America and in Europe during the first quarter of this century, I was struck by the domination of discussion by classical geologists at the expense of informed comment from geophysicists. For the first time I understood how geology, the science of the earth, came to be reduced in meaning to a part of the geosciences. It is quite clear why the geophysicists of the day joined with the geodesists to form a scientific union in 1919. It is interesting that it took another 41 years for geologists to decide that a geological congress every three or four years was not enough to make the disparate geological community reasonably cohesive. It was my good luck to be involved in developing the concept of the union and later in developing the union itself. It was a fascinating experience.

I came to the discussions about the union from the Geological Survey of Canada where there were fully integrated units of geologists, palaeontologists, geographers, geophysicists and geochemists. I would like to see much more direct contact between groups of earth scientists, of all those people who deal with the earth as a physical unity. Perhaps this amalgam will be helped by satellite technology, as it is now being helped by the complexity of marine sciences. Imagery will be very useful for many sciences, including geoscience, as will other data obtained by remote sensing. Ground truth is needed if the imagery is to be interpreted successfully in other areas, and if there is any place on the Earth where there is an abundance of geological data observed on the ground it is in Europe. The interpretation of satellite imagery over wide varieties of geological environments could surely be led by Europeans. Geophysical inputs are essential as well and to be effective geologists and geophysicists must work together. Now that geological data are being quantified in a way undreamed of 25 years ago, it is possible that the communication gap is not as wide. It will require all the skills of geoscientists to interpret the data in regions where 'ground truth' is scarce, and I strongly suspect that even in Europe such combined studies will show up weaknesses of interpretation that have been obscured by national boundaries. Since the imagery is not limited by boundaries, all earth scientists will have access to data to make it possible to question or to amplify previous interpretations. It will also help to expose the personal fiefdoms of some geologists to a broader public and result in some important localities being studied by a wider range of specialists.

I do not need to make the point here that geology is a fundamental science when problems of the environment are being considered. Nevertheless this aspect of the geosciences is especially important in Europe with its relatively dense population, advanced state of industrialization, and relatively low supply of mineral and fuel resources. Before intelligent planning can be undertaken there must be a solid foundation of information on which to base alternatives, for no matter how much we may dislike the idea, we will inevitably have to sacrifice some environmental quality for other presumed benefits. At the moment, evidence is scarce of attempts to preserve any environmental quality in Europe, although I am bound to admit that the improvement of the atmosphere in the London area is a splendid example of what can be accomplished with determination, and with money. Still, since much of the obvious pollution comes from the consumption of energy and from the utilization of mineral resources, geoscientists have increasing responsibilities for interpretation, for analysis and synthesis, and for prediction so that the public, the government and industry may better plan the utilization of materials, energy, mineral and the land surface itself. European geologists can integrate their work with that of geophysicists, geochemists, soil specialists, hydrologists etc. in a way that should lead the world. As environmental scientists they are in the forefront and can further improve their leadership role. In the meantime they have awesome responsibilities for ensuring the evaluation of resources for the long term, plus the exploration for resources in the immediate future. I suspect also they will have to keep a highly important segment of earth scientists engaged on the advancement of knowledge for its own sake because elsewhere in the world the great bulk of them are, understandably, engaged on highly pragmatic activities of immediate concern to their national needs.

How are they to do all these things? It is the custom everywhere to be highly critical of the existing structure if it does not produce what the masters think it should. So instead of examining the structure to see if it can be made to work, a new structure is imposed. I have been through a discouraging number of such reorganizations, most of which have resulted in little or no improvement. The organization chart should not be thrown away, it should simply not be inked in. Usually there is a sound basis for its structure, but it must be continually adapted to changing conditions. Already there is a plethora of geological groups—governmental, professional, academic, honorary, national, inter-govern-mental and non-governmental—and the problem is how to make them work for the benefit of the scientists, of science and of mankind in general.

All European countries have a national or state geological organization, usually a survey, that is preoccupied with the problems of keeping up to date the geological map of the country. Since a geological map is the expression not only of more or less objectively observed data, but also the subjective interpretation of data from within the country in relation to similar published data from adjacent countries, it is a synthesis of geological research. There is no such thing as routine geological mapping; or if it is being done, it is not being done by truly profes-sional geologists. Everything put on a geological map has been interpreted to a greater or lesser degree and it is for this reason that all critical areas, or type areas, should be, as I indicated earlier, 'studied to death'. Surely the national surveys could take a lead here. There is also a rather special problem that I think must afflict European surveys, or at least many of them. We are used to living in a time of expansionist economics, but for many countries in Europe the national geological surveys are at a steady state, if they are not actually declining. When continually recruiting young scientists it is easy to keep the team young and forward-looking. But when most of the staff are older than, say, fifty years the problem of keeping a fresh outlook is paramount. How then to prevent ossification? I am personally convinced that exchanges of personnel between in-dustry, university and government ought to be a matter for national policy and everything possible done to facilitate them. A new point of view in an organization, and a new environment for an individual, can do wonders for both. Is it too much to have pensions portable between nations, or between industry and government? When dealing with people aged over 45 this is a prime consideration.

Parallel with the surveys are the national geological societies. They are of great importance because of their influence on the development of geosciences on the national scale, and for their influence on international activities. Even though most are at least partly funded by governments many are nevertheless to a large degree independent of governments. They can greatly influence the direction taken by the nation in its development of science, and can have a strong effect on national participation in geological programmes. I am sure that scores of people who have come to Reading are members of state surveys, but I am sure that a meeting entitled 'European Geological Surveys' would be much more difficult to convene because governments would then have to be involved. Freedom of action is one of the great advantages of the scientific societies, and I hope the outcome of this meeting will result in specific steps for improved consultation between the countries of Europe. I also hope, however, that the consultations will not be limited to geological societies but will include the spectrum of geosciences.

I suggest that geologists could provide a useful service through meetings devoted to topics to which all geoscientists could contribute. They would then be meetings of European Geoscience Societies, or rather meetings of European geoscientists. Such meetings would help to focus attention on problems, rather than disciplines, and would help the younger geoscientists to become aware of the fact that there is more than one kind of interesting and useful tree in any forest. This is important but not enough.

At the risk, again, of appearing chauvinistic I would like to point out to you a Canadian experiment. Admittedly it is easier to do there with only two languages and nine provinces plus one federal government active in geosciences. A Geoscience Council has been formed of representatives of each of the specialized societies in geoscience. The Council reviews the state of things, attempts to forecast requirements for geoscientists, and plans to keep the appropriate governments better informed on problems facing the geoscientific community and the governments themselves. In the last sense they are of course, a pressure group or lobby. Amongst their activities is that of sponsoring earth science education in Canada, where each province has sole responsibility for its curriculum. In the 1974 report, the Council reports that the number of students in high school receiving identifiable earth science education is increasing rapidly in many provinces. It is considering other matters such as geodata systems, problems of funding research, and the like. Thus it is able to focus on the needs of the whole geoscience system. Basically however, it deals with 'matters of interdisciplinary communication and broad policy'. Perhaps something of that nature would be useful in Europe.

Recently the European Science Foundation has been formed, which I believe plans to develop large-scale programmes in science supported and carried out by joining the capabilities of the Research Councils. It is directing its activities to the western part of Europe and is non-governmental. A meeting has been called by Yugoslavia to consider establishing a European Network for Science, with the idea of stimulating particular science projects of interest to several countries in both parts of Europe. If it goes forward, it will be inter-governmental and presumably UNESCO will be asked to provide the secretariat. The Conference on European Security and Cooperation ended with what is generally known as the Helsinki Declaration which, amongst other things, calls on European states (including USA and Canada) to improve their co-operation in science and technology. If we are not careful, therefore, there will be so many people busy co-operating there will be none left to do the work. Still, I strongly support the concept of geosciences as a unit devoted to the study of the earth. It is a field of enormous importance for the years ahead and for advancement of all countries. I would like to see geologists take the lead in a re-unification and the time is now ripe, for governments are anxious to make the Helsinki Declaration something more than a pious hope. Good proposals will be welcomed.

One of the most important vehicles for such co-ordination is the International Union of Geological Sciences (IUGS). This international non-governmental organization, although inadequately financed, and therefore inadequately staffed, is doing a great deal to pull together the subdisciplines and works continually closer with the International Union of Geodesy and Geophysics (IUGG) especially in those fields that have substantial 'traditional' geological activities such as the International Association of Volcanology. Similarly, joint activities with the International Geographical Union (IGU) are developing, particularly in the studies of the Pleistocene and geomorphology. All this takes place under the umbrella of the International Council of Scientific Unions.

Although non-governmental, the influence on governments of IUGS and other scientific unions is great. We have only to think of the International Geophysical year, Global Atmospheric Research Programme, and now the International Geological Correlation Programme jointly launched by IUGS and UNESCO. It officially began in 1973 and already more than 80 nations are either participating or are planning to participate. Even though it is a global programme it has to be carried out by nations, commonly in a regional context, and I think the European region should lead. This is a wonderful opportunity for meaningful co-operative projects that will benefit the whole of geoscience. Because of its global approach, the bell may be tolling for the restrictive procedures concerning some areas where uninvited outsiders are not welcomed. Other signs though, are not so good. Too many projects are being proposed to the IGCP that are rather tired hobby horses of particular geologists. European geoscientists are the worst for this. Inevitably some become incorporated in the programme and there is a real danger that, unless the geological community puts forward bold and imaginative projects, the IGCP will never realize its potential. Surely a council of European geological societies could help to focus the programme and to ensure that projects submitted would be supported actively by this part of the world community.

The proposal for an IGCP was approved by the 17th session of the General Conference of UNESCO in November 1972, it having been approved by IUGS during the International Geological Congress in August 1972. Members of the IGCP board are appointed jointly by the Director General of UNESCO and by the President of the Union. Some people still believe UNESCO was not needed but I am satisfied now that, without UNESCO's participation, many countries now involved in the programme would be absent if their governments had not been committed through UNESCO. At any rate, it is an interesting experiment that depends on the geoscience community to make it work. Some government representatives at the last meeting of the General Conference stated that they thought the programme should be run entirely by UNESCO with only scientific support from the Union. Since the programme is scientific and is almost entirely covered by activities of the Union, geologists I have spoken to think it much more logical to have a joint programme, UNESCO's share being to provide the bulk of secretariat services, some financial support and considerable stimulation through its General Conference.

UNESCO is important for international science because programmes discussed and committed by the General Conference commit governments. Thus, non-governmental organizations can conceive and plan major programmes, and their adoption by an inter-governmental organization provides commitment. The IGCP, Man and Biosphere, Unisist, Inter-governmental Oceanographic Commission and the International Hydrological Programme are sponsored jointly by the international scientific community and by intergovernmental organizations. I am convinced that the international nongovernmental and inter-governmental agencies together are much more effective than the two working alone. I urge therefore that you, the geologists of Europe, play the lead role that is required. Use your influence to ensure that geological programmes are sound and are really significant. You are the people who can and must take the lead in giving direction to the global international non-governmental agencies, and you can exert your influence at home on the delegates that represent you at the various assemblies of the UN system. It is not by talking to each other that geoscientists will make the world community aware of their capabilities, and the critical role they can play in the years ahead. We must make it generally known.

Now, several people have asked me how I think this proposed European 'federation' is to work. It depends first, of course, on the will to make it work which in turn depends on its felt need. I think you need it and I hope this meeting ends with at least proposals on which action can be taken. Assuming that you do develop some form of federation, should it float freely or be attached to some international organization? I am inclined to think it should be basically free though it makes sense to me to consider affiliation with IUGS. Too close an association would, I think, be potentially disrupting for the Union. The Helsinki Declaration includes comments on the responsibilities of Europe towards the Third World, new international economic order, and the like. It is only natural that, at present, Europeans should dominate the scientific unions but this must and will change as capabilities increase in the developing countries. All scientific unions need to be seen by the societies in the developing world as important for them. A direct affiliation of a European 'federation' with any union might be taken in some quarters as further evidence that the unions are for the scientists of the industrialized world. Perhaps something like 'consultative status' might be considered. Of one thing I am certain: the 'federation' should be financially independent of the Union.

Should it affiliate with UNESCO? I think the same reasoning applies here, possibly with even more emphasis, for UNESCO's General Conference is made up of people appointed by government. However, we have a Bureau for European Scientific Co-operation which would certainly be a good point of contact. At the moment the Bureau consists of one individual, but we expect this to change soon and, especially following Helsinki, I would think that the possibility of establishing links with UNESCO through its Bureau are promising.

THE ROLE OF GEOSCIENCES IN MODERN SOCIETY

GERD LÜTTIG

Bundesanstalt fur Geowiddenschaften und Rohstoffe und des Niedersachsischen Landesamtes fur Bodenforschung, 3 Hannover-Buchholz, Stilleweg 2, Federal Republic of Germany

It is not hysteria but wise forecasting when futurologists, raw materials experts, nutrition scientists, environmental researchers and technologists emphasize the shortages, price increases and barriers in the natural environment towards which our generation, devoted to growth and progress, is heading. Humanity finds itself not only in an economic and sociological, but also an ethical, crisis. We geoscientists can evaluate the part of this crisis determined by the natural environment. We know that the situation is serious and that we must make every effort to extend the lifetimes of raw materials and other natural resources such as water and soil. We can do this by discovering new reserves, by finding substitutes for diminishing raw materials, by utilizing possibilities of recycling, by planning so as to avoid improper use or squandering of the potential of the environment, and by not polluting through waste disposal.

We apply our energies to finding scientific answers to such practical problems. However, solving these problems and carrying out the solutions we leave to the planners, futurologists, technologists, and engineers. *That is wrong!* In the sense of *prospective geoscience*, we must leave our ivory towers and engage actively and directly in planning. Geologists belong in every planning staff, in every institution involved with economic and environmental development and, also, in the relevant political bodies. They should be at the forefront everywhere that involves the course and security of the future of mankind. The European geological societies should see to it that this demand is publicly articulated, that appropriate scientists are 'sacrificed' to the public for the good of all geoscience, that future geoscientists are appropriately trained, and that national representatives agree upon, coordinate, and control the politics of research within the geosciences. What is necessary here is action, not words!

In order to give you a better indication of what I intend to bring forward, I ask you to enter into the feelings of one of your confrères of geology of my generation on some past day. Born in the '20s somewhere in central Europe, be it Budapest or Amsterdam, Berlin or Warsaw, he would have started his studies after World War II, full of the criticism of his brothers against authority, strongly affected by what happened just before, without any idea about the future and the possible role of a geologist in society. For what was society at that moment except a multitude of materially minded, living from hand to mouth. When his father asked him what his idea was with respect to a career, to the position from which he might gain a certain status as a member of academic society, our colleague would have answered that this was by no means the reason for taking up geological studies but that this happened for reasons of pure science, that he wished to look into the mysteries of nature, having the opportunity to be in contact with all the miracles of the natural world, in the mountains, by rivers and lakes, on sea-shores and in the wonderful deserts. There was almost no relation to practical daily work; but after some years of earnestly exploring all the details of this part of science, our colleague found that mankind, in order to overcome the difficulties of the post-war period, encouraged by new aims in life or pressed by the necessities of the situation, had some tasks for the geologist. And now he came into contact with an environment in which he could apply all his special knowledge and capability in a very favourable way.

In central Europe this was the time of urgent increase of agricultural production. Soil mapping for amelioration purposes, land reclamation, surburban development, was enormously increased. There was increased demand for ground-water supplies, coal and peat resources had to be studied, clay for brickworks had to be investigated for reconstruction of all the cities which had been destroyed. This was the time when the Geological Surveys and all other scientists capable of technical advice were like a fire brigade, offering an emergency service in many areas. That's why we very soon find our friend leaving his gastropods or rhinoceros bones and jumping into all the aspects of practical geology, where he gets a place as one of the (generally passive) collaborators in this emergency phase.

The second stage occurred after a few years, under the name of the economic miracle: a period of rapid economic expansion, in which our fire brigade was asked to look beyond national and European borders. No man with a flair for unravelling mysteries, with this specific touch we geologists have, will ever forget in his life this first moment when he crossed his own regional and mental boundaries, sniffing the first smell of rosemary on an Apennine slope, the unforgettable smell of burning shish kebab and the taste of ouzo on his tongue.

However, in this phase, which brought such remarkable knowledge to geology; in which the roots of modern marine geology are hidden; in which modern metallic and non-metallic mineral exploration methods were born with extensive geophysical and geochemical prospecting rounds; and in which we as European geologists had so many opportunities and so much success studying the whole world; in this period, lie the roots of all the evils we now have to overcome.

For this economic expansion has accelerated the demand for raw materials. Society's demands for goods have (often artificially) been encouraged by industry; we have been manipulated into being a community throwing away many things, producing steadily increasing amounts of waste. Exploitation is greedily destroying deposit after deposit. Suddenly we are to be found, to the astonishment and distress of the people responsible for mankind's development, face to face with shortcomings, cost rises and other economic enigma which seem insoluble.

This is the critical situation in which our fellow geologist finds himself. This is where I would like to hammer into every colleague's head that this should not only be a moment of contemplation but also of very active revision of views and reorientation of our geoscientific politics. This generation and the next occupy a very remarkable place in the history of mankind. Nowadays, ever-increasing numbers of intelligent people are realizing the limits that are set to the potential of our natural environment. We who are coming against these barriers find that, even in countries which to our fathers seemed to be lands of infinite opportunity, shortages of energy, raw materials and food, and environmental pollution, are all too familiar.

Insecurity is widespread. Symptoms of hysteria can easily be discerned in the international press. The anti-pollution movement engulfs even clear-thinking and cool-headed engineers and planners like a wave of apocalyptic medieval anxiety. On the other hand, some futurologists, energy and raw materials experts, like others, speak comfortingly to the man in the street and babble of inexhaustible riches.

The truth is that considerable unexploited environmental potential still exists. It is, however, also true that not just here and there, but over wide areas, we are approaching, by something like an exponential curve, limits of the environmental potential which it will be impossible to remove.

Contrary to the view of some futurologists who lay themselves open to the charge of lack of appreciation of the physical environment, earth scientists are bound to doubt even whether the next generation will survive. This doubt is due not only to the critical situation in many parts of the world, mainly Europe, with respect to raw materials and energy resources, but also to the scarcity of groundwater, land suitable for agricultural production, good construction ground, waste disposal facilities, etc.

What is the function of geology and related sciences in modern society? The important thing in this situation—this is the very heart of the matter—is the realization that the geoscientists investigating the potential of the environment have not only been mapping it for decades but also investigating the natural resources contained therein, and the possibilities of their practical utilization. Therefore they are predestined to help to find a way out of the difficulties I have talked about.

This is why I call for a reorientation of our research and its priorities. Geologists and geoscientists from neighbouring disciplines, especially those in Geological Surveys, should operate well in advance of the battlefield of futurologists, country planners, economists and similar experts who are wrongly considered by the public to be the sole purveyors of salvation.

Geoscience started as a branch of science looking *into the past;* Earth history is one of the basic fields of our subject. Earth's and mankind's *future* can be another and very important one. For this reason I would like to see established a newly orientated earth science, called 'prospective geology'. This future-orientated geology has nothing to do with supporting hysteria. Its task is to rouse humanity, which has no thought for anything except daily drudgery and is led by shortsighted profit-hungry managers, and also to dedicate itself to the service of mankind. Academic studies leading merely to the compilation of documents for industrial, agricultural, water-resource, or traffic planning are no longer good enough. Instead we must advance into the territory beyond the frontiers of economic policy planning so that we can identify the danger points, and exert direct influence on planning for the natural environment which is becoming increasingly restricted.

I therefore understand prospective, forward-looking geoscience as governed by a strategy designed to advance parts of geoscience beyond the role of a science which acts on request, frequently and unfortunately late, sometimes too late, in other cases only slightly ahead of the planners, and invited or uninvited, into the role of futuro-technology.

Is there not abundant evidence that entire prospecting campaigns for non-ferrous metal ores (for example, copper, gold, silver) were only begun at a moment when, predictably, the price of the raw material had markedly risen? Similarly with engineering projects—geological comment on civil engineering works was only proffered when shortcomings or defects in planning, foundations, or construction work had become evident. In the same way geological mapping was only concluded after industrial projects had been completed. And so on.

We should not have to seek remedies when symptoms are evident, but we should have a solution ready long before the crunch occurs. Don't say we are not in a position to act in such a way; we are among the few people who are.

Some of my professional colleagues will consider this to be mere fancy talk and go on to ask how this dream is to be realized. Realization is easier than it would seem, however, A few examples may be given:

(1) Raw material forecasts which are made by several bureaux and surveys show relatively exact static lifetimes and a good impression of the dynamic lifetime, so that after studying the figures one is able to evaluate development and the necessary exploration steps, regionally and with regard to the mineralogical and genetic type of deposits.

(2) This exercise gives a good start for calculating the necessary research in the fields of ore concentration, preparation, dressing, new methods of leaching, metallurgy, petrological, geochemical, and other modes of prospective geotechnology in deposits to be worked in the future.

(3) Recycling aspects and the search for substitutes will increasingly have to be examined in conjunction with problems of disposal of waste materials.

With respect to the search for substitutes I consider it wrong to leave the discovery of new materials mainly to the chemists as was done in the past, because it is the geoscientists who study and master the physical and chemical properties of minerals and rocks. Should it not be our responsibility to discover and recommend the use of a different mineral and rock, to replace a raw material, or rock, or ore, which is no longer available for some application? We should stop depositing bulk materials at the feet of the technologists, leaving them to discover the solution, whilst we withdraw to make maps, determine reserves and act as the traditional water-carrier.

(4) In the field of looking for new energy sources we do not have to restrict ourselves to geothermal research and the geology of oil shales or things like that. There will still be a lot of geological problems in the next century when we are able to use nuclear fusion and other, non-fossil, forms of energy. Even now we should start to train in all the technological disciplines that we need so as to play a role in advance of the problems.

(5) Even pure geognosy can be used to serve the interests of prospective geology.

We know that without geological mapping and the basic recording of geological data, any applied geologist would be as helpless as a miner or an engineer faced with a certain practical problem of geological origin.

Even from the purely geological map, dominated on the whole by lithostratigraphic information, other kinds of maps have been developed (ground-water, soil, foundation, ore deposits, oil and gas, raw material and other maps), which are excellent sources for modern natural environmental planners. However, if we geologists could slip out of our own self-understanding for one moment and function as objective observers in a dialogue between regional planners, we would realize that in most cases our documents require considerable interpretation before the planner can use them. In some cases we need interpreters, and in other cases we find very clever people earning a lot of money with the help of that translation.

In many cases we leave the planner alone with sets of geoscientific maps of an extraordinary academic standard. If he has a conflict situation in his planning, let us say between ground-water utilization, agriculture, foundation engineering and the winning of raw materials (for example, sand and gravel), and if every factor may be of advantage in the region concerned, we usually give him no help after having fed him with our maps, pretentiously named basic maps for planning. Who can help the planner in such a conflict case, if not the geologist? Isn't it the geologist who can give the correct proposal for the best land use?

Such considerations, prompted by the practical experience of planning-orientated geoscientists, have led to the suggestion of a new set of maps: *geoscientific maps of the natural environment potential* (GMNEP). This project has been taken up by the International Geological Map Commission of IUGS; a UNESCO Symposium about this type of map has been held at Hanover, in April–May 1975. The meeting of the IUGS working group will be this winter, and the Hanover Court of Audit has introduced this map as obligatory to all ministries concerned with planning in the Federal Republic.

These examples of possible changes of the geologist's role in society can be multiplied. The important thing, as should be pointed out to sceptical colleagues who intend to continue in their cultivation of purely academic research, is that our Earth is no longer an idyllic place where the geologist, as a perfectly educated man like the one I spoke about at the beginning, rises spotless above other citizens of this Earth. Today we are suddenly faced with barriers to development which can only be surmounted by the concerted action of all men of intelligence, wisdom and mutual consideration.

For that reason I ask a few of my colleagues—notwithstanding that basic scientific study, because of the enormous role it is playing in the development of science, must go on—to step out of their ivory towers and engage actively and directly in planning. For this purpose they have to learn the language and attitudes of planners, and we professors have to give our students an education suitable for this action.

In the future, geologists should belong to every planning staff of regional importance. There should be at least one as a member of every institution involved with economic and environmental development, and in every relevant political body. *They should be at the forefront wherever the course and security of the future of mankind is involved.*

The European geological societies should see to it that this demand is publicly articulated, and as the individual views of the different national societies may be too weak and insufficient, I herewith appeal to all scientists present, because it is the individual who makes up society, to decide upon the creation of a European Geological or Geoscientifical Society, here and now. Among other things this European Geological Association should see that appropriate scientists enter public service for the good of all geoscience, that future geoscientists are appropriately trained and that in the different nations every influence is used for co-ordinated and reasoned influence on all politics related to geoscience.

However, what is necessary here are not words but actions, and if one of those actions could be a petition of all scientists present for the creation of such a society, in whatever form may it be created, and of a board taking over the measures of organization, statute making and foundation, I would be the happiest geologist who ever came to Reading.

INDEX

Acadian phase, 111
Adriatic plate, 151–152
Adriatic Sea, 37
Aegean arc, 148
Aegean plate, 149, 151
Aegean Sea, 37–38, 148–149
Afghanistan, 34
Africa–Arabian platform (craton), 26, 31, 85, 91
Alberta, 46
Alboran zone, 37
Algeria, 20
Alps, 29–30, 33–34, 91, 98, 121, 143 et seq.
Anatolia, 31, 152
Anglo–Paris basin, 36
Antarctica, 178
Apennines, 150
Apulia, 150–152
Archidiskodon, 170
Ardennes, 82, 106
Ardennian disturbance, 111
Armenia, 37
Armorica (massif), 26–28, 31, 96, 121, 125–126, 129–130
Arzakanian massif, 33
Asia Minor, 152
Asturia, 31, 131
Attic–Cyclad massif, 152
aulacogens, 28, 30, 35, 36
Australia, 177
Azerbaijan, 170
Azov Sea, 37

Baie d'Audierne, 28
Balkans, 150
Baltic Sea, 23, 37, 94
Baltic shield, 35–36, 63, 66, 68, 70, 74
Barents Sea, 37
Barrandian basin (syncline), 97–98, 129
Bay of Biscay, 28, 37
Bay of Corinth–Menderes zone, 37
Bear Island, 37
Belamoride belt, 63
Belgium, 35
Benioff zones, 31, 34, 125, 132
Betic Cordillera, 32

Black Forest, 124
Black Sea, 37, 38, 102, 121
Blankenburg zone, 129
Bohemia (massif), 23, 29, 71, 97–98
Böllstein Odenwald, 126
Borna–Hainichen basin, 130
Brabant massif, 23, 105–106, 121
British Midlands, 23, 97
Brittany, 96, 102
Brunhes (geomagnetic) epoch, 107
Bukk mountains, 29, 32

Cabo–Ortegal–Barganca, 29
Calabria, 82
Caledonian front, 71, 76
California, 149
Canadian shield, 74
Cantabrian zone, 121, 127
Carnic Alps, 30, 32
Carpathians, 23, 29–30, 34, 61, 102, 152
Caspian (Sea), 36
Caucasus, 29, 32, 37, 150
Celtiberian aulacogen, 36
Celtic Cycle, 93–94, 102
Central French massif, 29, 94, 124, 130
Central German crystalline rise, 27, 121, 125, 129–130
Central Russian depression, 68, 70
Champtoceaux nappe, 31
Chateaulin–Laval syncline, 129–130
'Coal channel', 31, 130
continental drift, 162–163
convection, 44, 49, 51–53
'core paradox', 52–53
Corinth (Bay), 37
Corsica, 82
Crimea, 29, 32
Cyclades, 148, 152

Danish–Polish aulacogen, 36
'Danish triangle', 65, 75–76
Dardanelles, 38
Denmark, 84
Dinarides, 29, 31–32, 150
Dnieper–Donetz rift, 38
Dobrogea, 29
Domnonea (see also Armorica), 26–27, 96

Donetz(–Donbass) basin (trough), 35, 68, 71, 73–74
Dzirula massif, 31

Earth's magnetic field, 48
East Elbian massif, 24, 121
East European craton (platform) (*see also* Russian platform), 21, 23, 25–26, 83–85, 87, 91, 94, 96, 102, 121
East Germany, 102
East Lusitanian–Alcudian zone, 122
Ekne disturbance, 111
Elbe Valley syncline, 129–130
England, 121, 125, 128, 135
English Channel, 37
English Midlands, 23, 97
Epi-Baikalian platform, 23
Erzgebirge, 29, 124, 126, 129

Farallon plate, 49
Fen complex, 75
Fiji, 46
Finland, 64
Finnmark, 94, 97, 102, 110–111
France, 135
Frankenberg massif, 130
Fyn–Grinsted high, 23, 65

Galapagos Islands, 44
Galician–Castilian zone, 122, 126
Gavrovo zone, 150
geodynamo, 49, 52
geomagnetic field, 51
Georgia, 31, 170
Geoscientific maps of the natural environ-ment potential (GMNEP), 196
Germany, 121
Gilmend continental block, 34
Gondwana, 32
Great Britain, 81, 102, 111, 121
Greenland, 63, 74, 81, 84, 86, 88, 90–91, 94, 96, 108, 111–112
Grenville province, 65
Gulf of Bothnia, 37
Gulf of Finland, 37

Hagenfjord basin, 94
Harz mountains, 31, 98, 129, 136
Harzgerode zone, 129
Hawaii, 41, 44, 46
Hawaii–Emperor chain, 49–50
Hebrides (craton), 61, 63, 75
Hellenides, 31–32
Highland Boundary fault, 106

Hirschberg–Gefell anticline, 132
Holy Cross mountains (Swiety Krzyz mountains), 23, 103, 121
Horg disturbance, 111
Hörre–Kellerwald zone, 132
hot spots, 46, 77, 135
Hungary, 98, 103
Hunsrück mountains, 125

Iberia, 121, 126, 131
Iberian meseta, 26, 28–29, 36
Iberian–Aquitanian block, 29
Iceland, 44
Iltay nappe, 96, 108
Imandra–Varzuga downwarp, 35
Insubric line, 150
International Geological Correlation Programme, 190
Ionian Sea, 37
Iran, 21
Iranian plate, 151
Ireland, 97, 111, 121, 135
Irish Sea geanticline, 97

Jutland, 23

Kabylian massifs, 26
Kamsk–Kinelsk province, 75
Kanin–Timan range, 94
Kara Sea, 20
Karawanken, 30
Karelia, 64
Karelides, 61, 63–64, 66, 68, 70–72
Kattegat, 37
Kazakhstan, 21
Kazan–Sergiev aulacogen, 35
Kermadec, 41
Ketilidian massif, 63
Khibine Tundra complex, 73
Kola peninsula, 35, 73, 94
Krivoi Rog, 71
Kursk, 71

Ladoga (Lake), 37
Lake District, 97
Lapland, 63
large aperture seismic array (LASA), 41, 47
Lebanon, 82
Lewisian complex, 75
Limpopo belt, 63
Lizard complex, 132, 136
Lofoten–Vesterålen, 110

Macedonian trough, 149
magnetic reversals, 50, 53
Malmedy graben, 130
mantle, 43, 135
mantle–core boundary (MCB), 41–42
Manych zone, 36
Marmora (Sea of), 38
Massif Central, 29, 94, 124, 130
Menderes massif, 152
Mendocino fracture zone, 49
Mid-European depression, 23
Mid-German crystalline sill, 27, 121, 125, 129–130
Midland Valley of Scotland, 27, 35
Mjosa Lake, 35
Moho, 66–68, 76
Moine thrust, 106
Moldanubian zone (massif), 121, 124–126, 129–130
Moldanubicum, 71
Moldavia, 23, 37
Mont Blanc massif, 149
Montagne Noire, 31, 105, 128
Moravosilesian zone, 121, 125, 127
Morocco, 21
Moscow aulacogen, 35
Moscow aulacogen (rift), 35, 38
Moscow basin (syneclise), 35, 68
Münchberg gneiss massif, 126–127, 132
Murray River basin, 177

Nelson River, 64
New Caledonia, 152
New Guinea, 41, 152
Normandy (massif) (*see also* Armorica), 26–27, 96
North American plate, 47
North Anatolian fault, 37, 149–150
North German high, 121
North Sea, 35–37, 71, 73–76, 84, 96, 112, 121, 130, 178
Norway, 71, 91, 96, 110–111
Norwegian Sea, 37

Okhotsk (Sea of), 72
Old Red Sandstone continent, 74, 131–132
Onega (Lake), 37
ophiolites, 28, 31, 33–34, 149–150
Oslo, 36, 75, 96, 97, 99
Ossa–Morena zone, 122

Pachelma aulacogen, 35
Pacific plate, 46, 49

Pakistan, 34
Palaeo-Tethys, 32
Pannonian basin, 98, 150
Paris basin, 125
Pechenga downwarp, 35
Pechora basin, 65
perimediterranean chains, 147
plumes, 44, 135
Poland, 23, 103
Pomerania, 121
Pomorze, 23
Pomorze–Moldavian zone, 24
Portugal, 126
Pripiat–Donets aulacogen, 35
Pyrenees, 29, 37

Randtröge, 76
Rapakivi, 25, 65, 73
Reguibat massif, 26
Rheic ocean, 130
Rhenish Schiefergebirge, 125–126, 128, 132, 135
Rheno–Hercynian zone, 27, 31, 121, 125, 130–132, 135
Rhine graben, 35
Rhodope massif, 33
Rif, 32
rift systems of Europe, 36
Ringköbing–Fyn high, 23, 65
Romania, 23, 30, 96, 102
Rötliegende basin, 130
Rügen Island, 23, 121
Russian platform (*see also* East European platform), 25, 35, 37
Ryazan–Saratov lineament (trough), 70, 73

Saale trough, 130
Saar trough, 130
Sakmar zone, 34
San Andreas fault, 149
Sardinia, 82
Saronic Gulf, 148
Sauerland, 126
Savic, 37
Saxony, 127
Saxothuringian zone, 121, 125–126, 130, 132
Scandinavia, 81, 86, 88–91, 102
Scoresby Sund, 108
Scotland, 27, 35, 75, 91
Scottish Highlands, 97, 106
sea-floor spreading, 163
Serbian–Macedonian massif, 29–30, 103
Serpont massif, 106

Sestri–Voltaggio line (zone), 150, 154
Sevan Lake, 37–38
Siberia, 20
Sicily, 82
Sillon houiller, 31, 130
Skagerrak, 37, 71
Skellefte ore-field, 64
Söröy phase, 110
South Portuguese zone, 122, 127, 135
Southern Uplands, 97
Sowie Góry massif, 126
Spain, 82, 126, 128
Spitzbergen, 81, 90–91, 98, 108
St. Anne trough, 37
Stara Planina, 29
Stavelot–Venn anticline, 130
subduction zones, 31, 130–131, 148, 156
Subvariscan zone, 121, 125, 127, 131–132
Sudetes, 29, 82, 128, 131
Svalbardian phase, 111
Svecofennides, 61, 63–64
Sveco–Norwegian block (province), 26, 65
Swiety Krzyz mountains (Holy Cross
    mountains), 23, 103, 121

Taimyr, 20
Tanne zone, 129
Taunus mountains, 125
Teisseyre–Tornquist line, 23–24, 71,
    75–76
Tell, 32
Tethys, 32, 34, 135
Thuringia, 132
Timan, 65, 70, 94

Timan–Pechora platform, 23
Tokaj mountains, 103
tonalitic line, 150
Tonga Islands, 44
Tornquist line, 23–24, 71, 75–76
Transcaucasus, 31, 33
Transuralian region, 37
Trondheim, 98
Trondheim phase, 110
Tunisia, 21
Turgay trough, 21
Turkey, 21, 82
Turkish plate, 149, 151
Tyrrhenian arc, 148

Ukraine (massif), 30, 63, 66, 68, 70, 74
Uppsala Resolution, 179
Uralian Foredeep, 37
Uralian–Mongolian belt, 31
Uralides, 73
Urals, 21, 34, 37, 66, 74, 102

Vardar zone, 148, 150, 154
'Variscan front', 135
Variscan zones, 121–122
VASA, 46, 47
Voronezh Massif, 21, 68, 70, 74
Vosges, 124, 132

Wales, 97, 121
Welsh basin, 96, 98
West Asturian–Leonesian zone, 121
White Sea, 37, 66